PRESTRESSED CONCRETE-LINED PRESSURE TUNNELS

Towards Improved Safety and Economical Design

PRESTRESSED CONCRETE-LINED PRESSURE TUNNELS

Towards Improved Safety and Economical Design

DISSERTATION

Submitted in fulfillment of the requirements of
the Board for Doctorates of Delft University of Technology
and of the Academic Board of the UNESCO-IHE Institute for Water Education
for the Degree of DOCTOR
to be defended in public
on Wednesday, 22 April 2015 at 10:00 hours
in Delft, The Netherlands

by

Tuan Dobar Yos Firdaus SIMANJUNTAK
born in Pangkalan Susu, Indonesia

Bachelor of Science in Civil Engineering, Institut Teknologi Medan, Indonesia
Master of Science in Hydraulic Engineering, UNESCO-IHE, The Netherlands

This dissertation has been approved by the promotor and co-promotor:
Prof. dr. ir. A.E. Mynett
Dr. M. Marence

Members of the Awarding Committee:

Chairman	Rector Magnificus, Delft University of Technology
Vice-Chairman	Rector, UNESCO-IHE
Prof. dr. ir. A.E. Mynett	UNESCO-IHE/Delft University of Technology, Promotor
Dr. M. Marence	UNESCO-IHE, Co-Promotor
Dr. ir. D.J.M. Ngan-Tillard	Delft University of Technology
Prof. dr. ir. J.A. Roelvink	UNESCO-IHE/Delft University of Technology
Dr. R. Kohler	Verbund Hydro Power AG, Austria
Prof. dr. A.J. Schleiss	École Polytechnique Fédérale de Lausanne, Switzerland
Prof. dr. ir. G.S. Stelling	Delft University of Technology, reserve member

CRC Press/Balkema is an imprint of the Taylor & Francis Group, an informa business

Published by:
CRC Press/Balkema
PO Box 11320, 2301 EH Leiden, The Netherlands
e-mail: Pub.NL@taylorandfrancis.com
www.crcpress.com – www.taylorandfrancis.com

ISBN: 978-1-138-02853-1

Summary

At the global scale, nearly two billion people are still lacking reliable electricity supply. Hydropower can be a source of sustainable energy, provided that environmental considerations are taken into account and economic aspects of hydropower design are addressed. Pressure tunnels are relatively expensive constructions, particularly when steel linings are used. Concrete linings can be economically attractive; however, their applicability is limited by the low tensile strength of concrete.

Techniques to improve the bearing capacity of concrete tunnel linings have become one of the interesting topics in hydropower research. One of the techniques available is through prestressing the cast-in-place concrete lining by grouting the circumferential gap between the concrete lining and the rock mass with cement-based grout at high pressure. As a consequence, compressive stresses are induced in the lining. This is meant to offset tensile stresses and avoid tendency for longitudinal cracks to occur in the lining due to radial expansion during tunnel operation. Moreover, as the grout fills discontinuities in the rock mass and hardens, the permeability of the rock mass is reduced. This is favourable in view of reducing seepage.

In order to maintain the prestressing effects in the concrete lining, the rock mass has to be firm enough to take the grouting pressure. The grouting pressure, taking into account a certain safety factor, should remain below the smallest principal stress in the rock mass. Since the prestress in the concrete lining is produced by the support from the rock mass, this technique is also called the passive prestressing technique. A classical approach to determine the bearing capacity of such tunnels does exist; but, it is based on the theory of elasticity assuming impervious concrete.

Due to the fact that the rock mass in nature is non-elastic and concrete is a slightly pervious material, doubts were fostered by experiences with tunnel failures resulting in loss of energy production, extensive repairs, and even accidents. Record shows that some of the tunnel failures are associated with hydraulic jacking or fracturing. While the former is the opening of existing cracks in the rock mass, the latter is the event that produces fractures in a sound rock.

The overall objective of this research is to investigate the mechanical and hydraulic behaviour of pressure tunnels. By means of a two-dimensional finite element model, the load sharing between the rock mass and the concrete lining is explored.

This research deals with the effects of seepage on the bearing capacity of pre-stressed concrete-lined pressure tunnels. A new concept to assess the maximum internal water pressure is introduced. The second innovative aspect in this research is to explore the effects of the in-situ stress ratio in the rock mass on the concrete lining performance. The rock mass supporting the tunnel is distinguished based on whether it behaves as an elastic isotropic, elasto-plastic isotropic or elastic transversely isotropic material. In the final part, this research focuses on the cracking of concrete tunnel linings. A step-by-step calculation procedure is proposed so as to quickly quantify seepage and seepage pressure associated with longitudinal cracks, which is useful for taking measures regarding tunnel stability.

If the assumption of elastic isotropic rock mass is acceptable, this research suggests that the load-line diagram method should only be used if it can be guaranteed that no seepage flows into the rock mass. Otherwise, seepage cannot be neglected when determining the bearing capacity of prestressed concrete-lined pressure tunnels.

In cases of pressure tunnels embedded in an elasto-plastic isotropic rock mass, the Hoek-Brown failure criterion is applicable for investigating the behaviour of the rock mass. When pressure tunnels are constructed in an inherently anisotropic rock mass, the rock mass can be idealized as an elastic transversely isotopic material. Regarding the behaviour of the concrete lining, the combined Rankine-Von Mises yield criteria can be used. While the former controls the response of the concrete lining in tension, the latter in compression.

When dealing with a three-dimensional problem of tunnel excavation and eventually the load transferred to the support, the limitation of two-dimensional models can be solved by means of the convergence-confinement method. However, this is not the case when the in-situ stresses in the rock mass are non-uniform. In such cases, the simultaneous tunnel excavation and support installation is acceptable provided that the radial deformations at the shotcrete-concrete lining interface are reset to zero to avoid the lining being influenced by the previous deformations during prestressing.

It is evident that the load sharing between the rock mass and the lining determines the bearing capacity of prestressed concrete-lined pressure tunnels. Particularly in the lining, longitudinal cracks can occur along the weakest surface that is submitted to the smallest total stress in the rock mass. When pressure tunnels embedded in elasto-plastic isotropic rock mass, longitudinal cracks may occur at the sidewalls if the in-situ vertical stress is greater than the horizontal. If the in-situ horizontal stress is greater than the vertical, cracks will occur at the roof and invert.

When pressure tunnels are embedded in transversely isotropic rocks and the in-situ stresses are uniform, the locations of longitudinal cracks in the lining are influenced by the orientation of stratification planes. If the stratification planes are horizontal and the in-situ vertical stress is greater than the horizontal, cracks can occur at the sidewalls; whereas if the stratification planes are vertical and the in-situ horizontal stress is greater than the vertical, cracks can occur at the roof and invert. When the stratification planes are inclined and the in-situ stresses are non-uniform, longitudinal cracks will take place at the arcs of the lining, and their locations are influenced by the combined effects of the in-situ stress ratio and the orientation of stratification planes in the rock mass.

Since crack openings in the lining are difficult to control with the passive prestressing technique, it is essential to maintain the lining in a compressive state of stress during tunnel operation. The attractive design criteria for prestressed concrete-lined pressure tunnels are therefore: avoiding longitudinal cracks in the lining, limiting seepage into the rock mass, and ensuring the bearing capacity of the rock mass supporting the tunnel. All in all, this research demonstrates the applicability of a two-dimensional finite element model to investigate the mechanical and hydraulic behaviour of pressure tunnels. Remaining challenges are identified for further improvement of pressure tunnel modelling tools and techniques in the future.

Samenvatting

Meer dan twee miljard mensen op de wereld ontberen betrouwbare energievoorziening en waterkracht is een manier van duurzame energieopwekking die daarin kan voorzien. Dit betekent dat nauwe eisen moeten worden gesteld aan het economisch ontwerpen van waterkracht-centrales. Leidingsystemen, met name de valpijp, spelen hierbij een belangrijke rol. Deze zijn vaak uitgevoerd in staal wat relatief gezien dure onderdelen van de constructie zijn die hun weerslag vinden in de bouw- en onderhoudskosten alsmede in de duurzaamheid van de energievoorziening als geheel. Door deze leidingsystemen uit te voeren in voorgespannen beton kunnen mogelijk kosten worden bespaard. Een van de mogelijkheden daarbij is om de ruimte tussen de betonwand en de rotsmassa op te vullen met een groutmassa onder grote druk. Op die manier worden de trekspanningen in langsrichting gereduceerd en kunnen barsten in radiaalrichting worden voorkomen. Bovendien wordt de doorlatendheid nabij de leiding gereduceerd, wat lekkage kan voorkomen en de stabiliteit vergroot.

Om de voorspanningseffecten van de betonwand te behouden, dient de rotsmassa stevig genoeg te zijn om de groutdruk aan te kunnen. De groutdruk dient, met inachtneming van een bepaalde veiligheidsmarge, onder de kleinste primaire spanningen in de rotsmassa te blijven. Aangezien de voorspanning in de betonwand wordt verkregen door middel van ondersteuning van de rotsmassa, staat deze techniek ook wel bekend als een passieve voorspanningstechniek. Een klassieke benadering om het draagvermogen van dergelijke tunnels te bepalen bestaat weliswaar, maar deze is gebaseerd op de elasticiteitstheorie en gaat uit van ondoordringbaar beton.

Vanwege het feit dat de rotsmassa in de natuur niet-elastisch is en beton een enigszins doorlatend materiaal is, zijn er twijfels ontstaan naar aanleiding van ervaringen met het falen van tunnels (valpijpen), welke resulteerden in verlies van energieproductie, dure reparaties en zelfs ongelukken. In sommige gevallen is het falen van tunnels in verband gebracht met hydraulic jacking of hydraulic fracturing. De eerstgenoemde is het verder opengaan van bestaande scheuren in de rotsmassa en de laatste is een oorzaak voor het ontstaan van scheuren in de intacte rotsmassa.

Het doel van dit onderzoek is om na te gaan hoe leidingsystemen van voorgespannen beton zich gedragen. Door gebruik te maken van tweedimensionaal eindige elementen berekeningen worden de krachten bepaald die hierbij een rol spelen. Het eerste deel van het onderzoek gaat na welke processen het draagvermogen beïnvloeden en lekkage veroorzaken. Daarbij wordt een nieuw concept geïntroduceerd om de interne waterdruk te bepalen.

Een tweede innovatie in dit onderzoek richt zich op het beter berekenen van het draagvermogen van de constructie door elasto-plastisch gedrag van de rotsmassa en effecten van anisotropie na te gaan. Met name dit laatste vraagt om een betere beschrijving van stratificatie-effecten in de omringende rotsmassa. Het laatste deel van dit onderzoek richt zich op het proces van scheurvorming in de betonwanden. Er wordt een eenvoudige methode voorgesteld om de lekkage vast te stellen en te kwantificeren zodat maatregelen kunnen worden genomen om de veiligheid en stabiliteit van de tunnel te garanderen.

Als de aanname van een elastisch isotrope rotsmassa acceptabel is, dan geeft dit onderzoek aan dat de load-line diagram methode alleen gebruikt moet worden indien gegarandeerd kan worden dat er geen lekkage plaats vindt in de rotsmassa. Indien dit niet gegarandeerd kan worden, dan kan de lekkage niet genegeerd worden bij het bepalen van het draagvermogen van voorgespannen betontunnels.

In geval voorgespannen betontunnels geacht worden ingebed te zijn in een elasto-plastische isotrope rotsmassa, is het Hoek-Brown faalcriterium van toepassing om het gedrag van de rotsmassa te onderzoeken. Als betontunnels gebouwd worden in een anisotrope rotsmassa, dan kan de rotsmassa voorgesteld worden als een, in dwarsrichting isotroop, elastisch materiaal. Met betrekking tot het gedrag van de betonwanden kan het gecombineerde Rankine-Von Mises criteria aangehouden worden. Waar de eerste de reactie van de betonwanden onder trekspanning controleert, controleert de laatste de compressie.

In het geval van een driedimensionaal probleem van tunneluitgraving waarbij de belasting wordt overgedragen op de ondersteuning, kunnen de beperkingen van een tweedimensionaal numeriek model verholpen worden door middel van de convergence-confinement methode. Echter, dit is niet het geval als de in-situ belastingen in de rotsmassa niet-uniform zijn. In zulke gevallen is het tegelijkertijd uitgraven en ondersteunen van de tunnel alleen acceptabel indien de radiale deformaties van de betonwanden verwaarloosbaar kunnen worden geacht, om te voorkomen dat de betonwanden beïnvloed worden door voorgaande deformaties tijdens het voorspannen.

Het is evident dat de verdeling van de belasting tussen de rotsmassa en de betonwand het draagvermogen bepaalt van de voorgespannen betontunnels. Vooral in betonwanden kunnen scheuren in de langsrichting ontstaan op plaatsen waar het zwakste oppervlak wordt blootgesteld aan de kleinste totale druk in de rotsmassa. Wanneer tunnels ingebed zijn in elasto-plastische isotrope rotsmassa's, kunnen scheuren in de langsrichting ontstaan in de zijwanden zodra de verticale in-situ belasting groter is dan de horizontale belasting. Wanneer de horizontale in-situ belasting groter is dan de verticale, ontstaan scheuren in het dak en de vloer van de tunnel.

Wanneer leidingsystemen ingebed zijn in dwarsrichting isotrope rotsmassa's en de in-situ belastingen zijn uniform, worden scheuren in de langsrichting in de wanden beïnvloed door de oriëntatie van stratificatie in de omringende rotsmassa. Als de stratificatie horizontaal is, en de verticale in-situ belasting groter is dan de horizontale, kunnen scheuren ontstaan in de zijwanden; waar als de stratificatie verticaal is en de horizontale in-situ belasting groter is dan de verticale, ontstaan scheuren in het dak en de vloer van de tunnel. Wanneer de stratificatie gekanteld is en de in-situ belastingen non-uniform zijn, ontstaan scheuren in de langsrichting in de bogen van de betonwanden en hun locaties zijn beïnvloed door de gecombineerde effecten van in-situ belastingen en de stratificatie in de omringende rotsmassa. Vandaar dat er vaak als ontwerpcriterium naar wordt gestreefd om scheurvorming geheel te voorkomen, lekkage te beperken en zorg te dragen dat het draagvermogen van de rotsmassa de tunnel ondersteunt. Dit proefschrift toont de toepasbaarheid van een tweedimensionaal eindige elementen model aan om het mechanische en hydraulische gedrag van voorgespannen betontunnels te onderzoeken. Uiteraard blijven er verdere verbeteringen mogelijk, zoals aangegeven in dit proefschrift.

Contents

1 | General Introduction

1.1. Background

A pressure tunnel in general is an underground excavation aligned along an axis and conveys high pressurized water from one reservoir to another reservoir or to turbine. As one of the hydropower components, pressure tunnels represent an important share of the total investment for hydropower plant. Without doubt, concrete linings have nowadays become the most attractive type of lining in view of construction time and economic benefits. Nevertheless, such linings are vulnerable to cracking during tunnel operation due to the low tensile strength of concrete.

By injecting the circumferential gap between the concrete lining and the rock mass with grout at high pressure, the bearing capacity of concrete-lined pressure tunnels can be improved. This technique, which is also known as the passive prestressing technique, can produce adequate compressive stresses in the lining to suppress tensile stresses and to avoid the opening of longitudinal cracks.

Principally, the lining prestressing is executed after the completion of consolidation grouting. This is necessary in order to provide stability to the underground opening after the tunnel excavation. Regarding the prestressing, the level of grouting pressure injected into the gap has to remain below the smallest principal stress in the rock mass. A full contact between the concrete lining and the rock mass can be achieved as the grout fills the gap and hardens. This provides a continuous load transfer between the lining and the rock mass, which is favourable for tunnel stability. Other benefits of this technique include homogenization of material behaviour and eventually stress pattern around the tunnel, and reduction of seepage into the rock mass.

Despite its popularity, the achievement of the passive prestressing technique depends on the characteristics of the rock mass. Due to fissures and discontinuities, the rock mass is obviously pervious. Even uncracked concrete linings are not totally impervious as often assumed by tunnel designers. Pores in concrete permit seepage pressures that act not only in the lining but also in the rock mass. Seepage pressures in the rock mass can affect the tunnel deformations and therefore should not be neglected.

Aside from taking into account seepage effects, the main novelty of this research is the determination of the load sharing between the rock mass and the lining, which has yet to be understood particularly when assessing the maximum internal water pressure. In view of the applicability of the finite element method in dealing with complex concrete and rock problems, finite element models can be used to address this task. Nevertheless, regardless of simplifications, analytical solutions should not be overlooked as they reflect both tunnelling tradition and design experience. A contribution towards an effective application of a two-dimensional finite element model on the design of concrete-lined pressure tunnels is presented in this dissertation.

1.2. Research Questions

In overall, this dissertation covers a series of investigations on the mechanical and hydraulic behaviour of prestressed concrete-lined pressure tunnels. It focuses on a deep, circular and straight ahead tunnel, which allows the application of plane strain, two-dimensional finite element models.

As the worst scenario for prestressed concrete-lined pressure tunnels, this research is dedicated for cases where tunnels are situated above the groundwater level. Without the groundwater, the bearing capacity of concrete-lined pressure tunnels depends solely on the prestressing works and the support from the rock mass. Distinction is made based on whether the rock mass behaves as an elasto-plastic isotropic or elastic anisotropic material. As the main research topics, the following research questions arise:

- What is the influence of lining permeability and the rationale to assess the maximum internal water pressure for prestressed-concrete-lined pressure tunnels?

- How different is the behaviour of pressure tunnels embedded in an elasto-plastic isotropic rock mass subjected to non-uniform in-situ stresses compared to those subjected to uniform in-situ stresses? Which parameter governs the tunnel bearing capacity?

- In cases of transversely isotropic rocks, how does the interplay between the in-situ stress ratio and the orientation of transverse isotropy affect the lining performance? Where are the potential locations of longitudinal cracks in the lining?

- Once longitudinal cracks occur in the lining, what is the procedure to estimate seepage associated with cracks around the tunnel? How does the saturated zone develop after the lining cracking?

- When using two-dimensional finite element models, what are the most important aspects for modelling of pressure tunnels? Which process is not considered in the model and affects the accuracy?

1.3. Research Objectives

This research aims to provide insights into how to determine the bearing capacity of prestressed concrete-lined pressure tunnels. Specific objectives are:

1. to develop a concept to assess the maximum internal water pressure of prestressed concrete-lined pressure tunnels and at the same time to quantify the amount of seepage into the rock mass;

2. to extend the applicability of two-dimensional finite element models to reveal stresses and deformations around the tunnel as a result of tunnelling construction processes, that consists of tunnel excavation, installation of support, and lining prestressing as well as of the activation of internal water pressure;

3. to identify potential locations where longitudinal cracks can occur in the concrete lining and introduce a procedure to estimate the seepage associated with cracks as well as its reach into the rock mass;

4. to derive the design criteria for prestressed concrete-lined pressure tunnels.

1.4. Dissertation Outline

Each chapter of this dissertation is written as a standalone article. In each chapter, a general background of prestressed concrete-lined pressure tunnels may be repeated, however with different emphases depending on the topic discussed. The five main chapters are Chapter 3, 4, 5, 6 and 7. While Chapter 3, 4, 5, and 7 already have been published elsewhere, Chapter 6 is under review and consideration for publication as another research paper.

Chapter 1 introduces the scope of this research. In addition to research questions, the specific objectives are listed. The outline of the dissertation is presented with an overview of content and structure.

Chapter 2 summarizes the state-of-the-art review of the design of pressure tunnels. Starting with the flow chart to quickly determine the types of pressure tunnel linings, historical development of prestressed concrete-lined pressure tunnels is presented. Furthermore, aspects in the design of prestressed concrete-lined pressure tunnels are highlighted. The knowledge gaps are identified.

Chapter 3 introduces the method to determine the bearing capacity of prestressed concrete-lined pressure tunnels in an elastic isotropic rock mass. Existing formulae to assess the prestress- and seepage- induced hoop strains in the final lining are recalled. A new criterion to assess the maximum internal water pressure is introduced. The effects of grouted zone on the stability of pressure tunnels are explored.

Chapter 4 investigates the behaviour of prestressed concrete-lined pressure tunnels in an elasto-plastic isotropic rock mass subjected to uniform in-situ stresses. It covers the modelling of tunnel excavation, support installation, prestressing of final lining and the activation of internal water pressure. Special attention is given to overcome the limitation of two-dimensional models when dealing with a three-dimensional problem of tunnel excavation. In view of model validation, the numerical results are compared to the available theory.

Chapter 5 further investigates the behaviour of prestressed concrete-lined pressure tunnels in an elasto-plastic isotropic rock mass. However, the tunnels being examined are embedded in the rock mass whose in-situ stresses are different in the vertical and horizontal direction. Two cases are analysed, based on whether the in-situ vertical stress is greater than the horizontal, or not. Locations where longitudinal cracks can occur in the final lining are identified, which is useful for taking measures regarding tunnel tightness and stability.

Chapter 6 deals with the behaviour of prestressed concrete-lined pressure tunnels in elastic transversely isotropic rocks. It explores the interplay between the orientation of stratification planes and the in-situ stress ratio, which is frequently ignored in the design of pressure tunnels. As well as potential locations of longitudinal cracks in the final lining, this chapter investigates the effect of anisotropic rock mass permeability on the saturated zone around the tunnel.

Chapter 7 focuses on cracking in pressure tunnel concrete linings. The concept to assess the internal water pressure resulting in longitudinal cracks is oriented towards the optimum utilization of the tensile strength of concrete. A simple approach to quantify seepage and seepage pressures associated with longitudinal cracks is introduced. However, numerical models are needed so as to capture the saturated zone in the rock mass as a result of lining cracking.

Chapter 8 summarizes the main findings of the research, arriving at conclusions and discussing remaining challenges and future works.

2 | Literature Review

This chapter briefly presents the historical development of the design of prestressed concrete-lined pressure tunnels. It provides a flow chart for an easy identification of the types of tunnel linings as well as the existing design approaches. The important aspects in the design of prestressed concrete-lined pressure tunnels are outlined. The gaps of knowledge, which need to be addressed in this dissertation, are identified.

2.1. General Design Criteria

The types of pressure tunnel linings in general depend on the characteristics of the rock mass covering the tunnel and the groundwater conditions. As a result, pressure tunnels may not be uniform in construction, but consist of different types of linings over their entire length. Fig. 2.1 shows the flow chart to allow for a quick determination of the types of pressure tunnel linings.

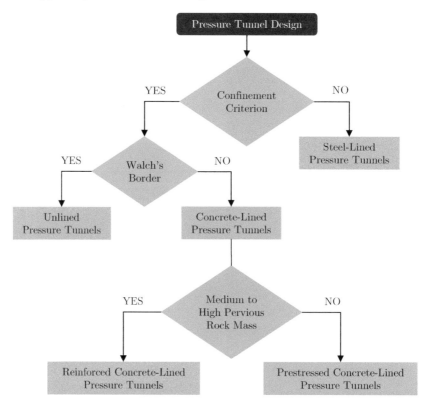

Fig. 2.1. Flowchart to Determine the Types of Pressure Tunnel Linings

As long as the smallest principal stress in the rock mass is higher than the internal water pressure, a steel lining is not necessary. Instead, pressure tunnels can be left unlined or merely lined with shotcrete for stability purposes if the rock mass is impervious and the external water pressure induced by the groundwater is higher than the internal water pressure. While the former criterion is known as the confinement criterion, the latter is called the Walch's border (Stini, 1950).

When the confinement criterion is satisfied but the Walch's border is not, concrete linings can be installed onto the shotcrete or the rock mass as an alternative to steel linings. However, the applicability of concrete-lined pressure tunnels is limited due to the low tensile strength of concrete.

Depending on the permeability of the rock mass, the bearing capacity of concrete linings against tensile stresses can be improved. If the rock mass is too pervious when compared to the concrete lining, an economical steel reinforcement can be embedded in the lining in addition to avoiding the occurrence of single wide cracks (Schleiss, 1997b). Like in most reinforced concrete structures, steel reinforcement in the lining can provide assurance against cracking. It distributes longitudinal cracks in the lining in a controlled manner.

If the rock mass is not too pervious, a carefully prestressed concrete lining can be adequate. A concrete lining can be prestressed either by grouting the circumferential gap between the lining and the rock mass at high pressure (Seeber, 1984; 1985a; 1985b) referred to as the passive prestressing technique, or by using individual tendons running in or around the concrete lining (Matt et al., 1978) known as the active prestressing technique. It has to be emphasized that prestressed concrete linings are not impervious. These types of linings allow seepage into the rock mass, which can influence tunnel deformations.

2.2. Historical Development of Prestressed Concrete-Lined Pressure Tunnels

The design method of prestressed concrete-lined pressure tunnels was first introduced by Kieser (1960). He introduced the so-called Kernring (core ring) lining as a substitute for steel linings. His method is characterized by the fact that the circumferential gap between the core ring and the rock mass is grouted with cement mortar which sets under pressure. The effect of prestressing in the core ring can be quantified by using the thick-walled cylinder theory (Timoshenko et al., 1970).

Thereafter, Lauffer and Seeber (1961) introduced the Tiroler Wasserkraftwerke AG (TIWAG) gap grouting method. Similar to the Kieser method, the concrete lining is prestressed against the rock mass by injecting cement-based grout at high pressure into the circumferential gap between the rock and the concrete lining.

In the gap grouting method of TIWAG, the grout is injected through the circumferential and axial pipes. These pipes, which are perforated, have valves and are placed at defined intervals along the tunnel wall before concreting the lining. As a result, the grout is more precisely distributed and an overall grouting of the circumferential gap between the rock and the lining can be guaranteed. As soon as the desired compressive stress in the concrete lining is obtained, the next pipe is connected to the pump. Another advantage of such arrangements is that the injection can be repeated as many times as required.

To facilitate the opening of the circumferential gap between the concrete lining and the shotcrete, the shotcrete surface can be covered with a bond breaker of whitewash or, a synthetic foil before concreting the final lining. Thereby, the grout will deposit in the circumferential gap and at the same time penetrates and seals fissures of the adjacent rock mass as the grout hardens.

In the Kieser and TIWAG method, the prestress in the concrete lining is produced by injecting the circumferential gap between the core ring and the supporting rock with cement-based grout. Since the compressive stress induced in the concrete lining depends on the support from the rock mass, this technique is known as the passive prestressing technique.

If the stability of the rock mass can be provided after the tunnel excavation and the installation of support, the lining can be prestressed. The concept of prestressing can be oriented towards the maximum possible utilization of the support from the rock mass. It relies on the stiffness of the rock mass to limit the lining deformations to the amount where no tensile stresses occur in the lining during tunnel operation. A necessary condition to maintain the prestress in the lining is therefore an adequate rock strength or rock overburden.

The successful application of the gap grouting method of TIWAG is mentioned at the Kaunertal power plant in Austria (Lauffer, 1968). In the 1980s, the gap grouting method was employed in the Drakensberg pumped storage in South Africa (Seeber, 1982; Sharp and Gonano, 1982). So far, the prestressing technique has been applied to many pressure tunnels around the world. It gains popularity since prestressed concrete linings are only slightly permeable and can be 30% cheaper than the use steel linings (Deere and Lombardi, 1989).

Other reasons for the popularity of the passive prestressing technique are continuous load transfer between the lining and the rock mass, reduction of the rock mass permeability and homogenization of materials around the tunnel. Recent publications are found in Wannenmacher et al. (2012) and in Grunicke and Ristić (2012), where this technique has been implemented to the Niagara Facility Tunnel Project (NFTP) in Canada.

2.3. Existing Design Approaches

2.3.1. Analytical Approach

Using the passive prestressing technique, the concrete lining and the rock mass are a composite construction. As a result, the load sharing between the rock mass and the lining can be calculated based on the compatibility condition of deformations. This is done by putting equal radial deformations at the boundary between the lining and the rock mass. Assuming elastic behaviour for both the concrete lining and the rock mass, Kieser (1960) employed the thick-walled cylinder theory to assess the internal water pressure. Thereafter, utilizing the support from the rock mass to the level of the smallest principal stress, Seeber (1984; 1985a) introduced the load-line diagram (Fig. 2.2), which is also known as the Seeber diagram method.

The load-line diagram is a method based on the condition in which the modulus of deformation of the rock mass within the range of stresses is known. It consists of the deformation line representing the load acting at the interface between the rock mass and the lining as a function of circumferential expansion.

The circumferential stresses at the interface between the rock mass and the lining can be expressed in terms of hoop strains. The slope of the deformation line is governed by the quality of concrete lining. The stiffer the concrete lining, the steeper the deformation line will become towards horizontal, and the lower the internal water pressure can be applied.

In view of the high compressive strength of concrete, the pressure injected into the circumferential gap can be high. Nevertheless, it should remain below the smallest principle stress in the rock mass so as to avoid hydraulic jacking or fracturing of the adjacent rock mass.

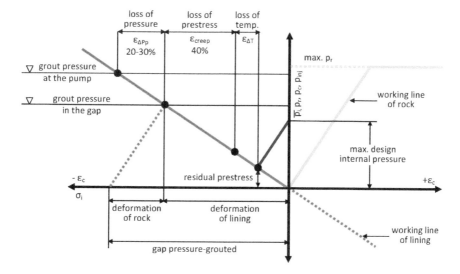

Fig. 2.2. The Seeber Diagram (Seeber, 1985a)

Until now, the load-line diagram method has been applied to determine the internal water pressure taking into account the loss of grouting pressure in the pump and prestress in the lining as a result of creep, shrinkage and temperature changes. This method is exclusively dedicated to a straight ahead circular tunnel embedded in elastic isotropic rock mass whose in-situ stresses are uniform. Especially in Austria, this method remains widely referred to. The most recent publications can be found in Marence and Oberladstätter (2005), Vigl and Gerstner (2009), and Wannenmacher et al. (2012).

Despite its popularity, the effect of seepage is not considered in the load-line diagram method. Concrete linings are assumed impervious, which is only true if waterproofing measures are employed. Without waterproofing measures, concrete linings are pervious due to pores in concrete. Furthermore, construction joints and fissures in the lining caused by shrinkage or cooling may permit seepage flow into the rock mass.

Since concrete lining is pervious, water will infiltrate cavities in the rock mass and develop seepage pressures. Seepage pressures affect rock deformations and can wash out the joint fillings in the rock mass.

In many occasions, severe seepage problems have caused not only the safety risk of the tunnel but also the loss of water and energy production (Deere and Lombardi, 1989; Panthi and Nilsen, 2010). Therefore, seepage effects cannot be ignored. When designing concrete-lined pressure tunnels, the mechanical-hydraulic coupling needs to be considered.

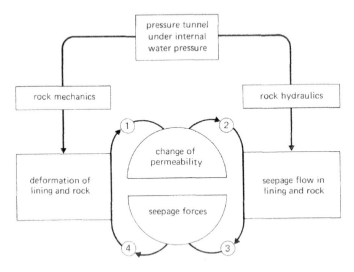

Fig. 2.3. Mechanical-Hydraulic Coupling (Schleiss, 1986b)

According to Schleiss (1986b), the mechanical-hydraulic coupling can be described as follows: the fractures and pores in the rock mass are deformed by forces so that the permeability in the rock mass around the tunnel is changed by the internal water pressure. In turn, the change in rock mass permeability affects the seepage flow and therefore, the seepage pressures.

The quantitative influence of seepage pressures on pressure tunnels can be estimated (Zienkiewicz, 1958; Bouvard and Pinto, 1969). In 1986, Schleiss introduced a method to quantify seepage out of pervious pressure tunnels, based on the porous thick-walled cylinder theory. He considered the mechanical-hydraulic coupling (Fig. 2.3) and emphasized that neglecting seepage pressures can result in an underestimation of stresses in the rock mass. Therefore, an accurate prediction of internal water pressure and seepage plays an important role in preserving the safety of prestressed concrete-lined pressure tunnels.

2.3.2. Numerical Approach

Thanks to the competence of the finite element method in dealing with geotechnical problems including non-linear deformability, material inhomogeneity and complex boundary conditions (Jing and Hudson, 2002; Jing, 2003), finite element codes have been widely used in rock mechanics and tunnelling applications. A detailed description of the finite element method is available in Zienkiewicz and Morice (1971) and Bathe (1982).

Revealing stresses and deformations as a result of the tunnelling construction process such as tunnel excavation, installation of support systems, installation of final lining and lining prestressing, and the activation of internal water pressure, is a challenging task. It requires the application of different material behaviour, such as concrete and rock. In particular, since the bearing capacity of prestressed concrete-lined pressure tunnels depends on the support from the rock mass, it is important to understand the failure of the rock mass itself. Also, pressure tunnels may be built not only in an isotropic rock mass, but also in an inherently anisotropic rock mass.

In cases of straight ahead circular tunnels and if one of the principal stress components is parallel to the tunnel axis, a two-dimensional plane strain finite element model can be adequate. Especially for tunnels situated above the groundwater level, numerical studies have been devoted to investigate the response of the rock mass to tunnel excavation considering a variety of rock mass behaviour, such as elastic isotropic (Stematiu et al., 1982), elasto-plastic isotropic (Swoboda et al., 1993; Wang, 1996; Carranza-Torres and Fairhurst, 1999; Carranza-Torres, 2004; Clausen and Damkilde, 2008; Serrano et al., 2011) and elastic cross anisotropic or transversely isotropic rock mass (Tonon and Amadei, 2003; Vu et al., 2013).

Regarding the numerical analyses of rock-support interaction, publications includes Einstein and Schwartz (1979) and González-Nicieza et al. (2008) when the rock mass is assumed as an elastic isotropic material, Carranza-Torres and Fairhurst (2000a; 2000b), Panet et al. (2001), Oreste (2003) when the rock mass behaves as an elasto-plastic isotropic material, and Bobet (2011) when the anisotropic rock mass can be idealized as an elastic transversely isotropic material.

The above mentioned numerical studies have contributed to the determination of excavation-induced stresses and deformations and the design of support. Nevertheless, a limited number of publications are found in the literature dealing with the design of hydropower tunnels. Particularly for the prestressed concrete-lined pressure tunnels, a few of them can only be found in Stematiu et al. (1982) and Marence (1996). While Stematiu et al. (1982) assumed the rock mass supporting the pressure tunnel as an elastic isotropic material, Marence (1996) considered the rock mass as an elasto-plastic isotropic material using the linear Mohr-Coulomb law.

2.4. Gap of Knowledge

If the assumption of elastic isotropic rock mass is acceptable, analytical solutions based on the elastic theory to assess the prestress-induced hoop strains and seepage-induced hoop strains are available. Yet, a criterion to determine the bearing capacity of prestressed concrete-lined pressure tunnels is still missing.

In nature, the rock mass is neither elastic nor isotropic. It may deform non-elastically as a result of tunnel excavation. Also, the rock mass does not possess a linear behaviour (Hudson and Harrison, 2001) since its strength depends on the principal stresses in a non-linear manner (Clausen and Damkilde, 2008). Therefore, when investigating the response of the rock mass to tunnelling, it is important to consider the non-linear Hoek-Brown failure criterion.

Predicting the load transferred to the support by the rock mass requires a specific approach that takes into account a three-dimensional effect of excavation. In this regard, two-dimensional models can still be attractive as long as the stress relaxation coefficient to account for the real delay of support installation is known. This can be solved by means of the convergence-confinement method (Panet and Guenot, 1982; Carranza-Torres and Fairhurst, 2000a; Panet et al., 2001). Nevertheless, this solution is applicable for cases of circular tunnels embedded in rock masses whose in-situ stresses are uniform.

Pressure tunnels may be constructed in an inherently anisotropic rock mass that are composed of lamination of intact rocks. Such a rock mass, commonly configured by one direction of lamination perpendicular to the direction of deposition, can take the form of cross anisotropy or transverse isotropy that exhibits significant anisotropy in deformability and permeability. Also, the in-situ stresses in the rock mass generally have different magnitudes in the vertical and horizontal direction. The interplay between transverse isotropy and the in-situ stress ratio and how these two issues affect the lining performance has not yet been studied until now and is frequently ignored in the design of pressure tunnels.

It has to be acknowledged that concrete linings are vulnerable to longitudinal cracks when loaded by high internal water pressure during tunnel operation. Once the lining is cracked, high local seepage takes place around the crack openings and can wash out the joint fillings that already settled in the rock mass. As long as the rock mass safety against hydraulic jacking or fracturing is ensured, seepage through cracks will produce losses and can be quantified analytically. Nevertheless, numerical models are needed to identify potential locations of longitudinal cracks in the lining as well as to assess the saturated zone around the pressure tunnel for tunnel safety purposes.

3 | The Gap Grouting Method[1]

This research confirms that using the sole load-line diagram method without taking into account seepage effects will result in overestimation of internal water pressure. In this chapter, a new concept to determine the bearing capacity of prestressed concrete-lined pressure tunnels embedded in an elastic isotropic rock mass is introduced. The maximum internal water pressure is assessed by offsetting the seepage-induced hoop strain at the lining intrados against the prestress-induced hoop strain.

Seepage into the rock mass has to be limited in view of tunnel safety. One of the remedial works is by grouting the rock mass. Aside from the new concept to assess the maximum internal water pressure, this chapter also discusses the role of grouted zone in improving the safety of prestressed concrete-lined pressure tunnels.

---

[1] Based on Simanjuntak, T.D.Y.F., Marence, M., Mynett, A.E. (2012). *Towards Improved Safety and Economical Design of Pressure Tunnels*. ITA-AITES World Tunnel Congress & 38th General Assembly (WTC 2012), Bangkok, Thailand. ISBN 978-974-7197-78-5.

3.1. Introduction

As long as the rock mass can be treated as an elastic isotropic material, the bearing capacity of prestressed concrete-lined pressure tunnels can be determined by using the load-line diagram method (Seeber, 1985a; 1985b), which was developed with the assumption of impervious concrete. In fact, concrete is a slightly permeable material and thus permits seepage into the rock mass.

If the rock overburden is adequate and the rock is of good quality, provided that the elastic modulus of the rock is at least one-third of that of the lining (Schleiss, 1986b), the long-term stability of concrete-lined pressure tunnels can be ensured by injecting the circumferential gap between the final lining and the shotcrete at high pressure with cement-based grout (Fig. 3.1). Since the prestress in the final lining is produced by the support from the surrounding rock mass, this technique is called the passive prestressing technique.

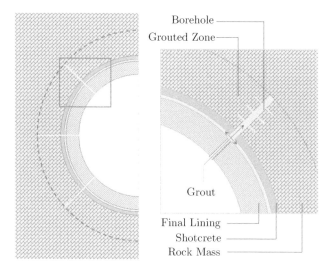

Borehole

Grouted Zone

Grout

Final Lining

Shotcrete

Rock Mass

Fig. 3.1. Schematic Geometry of a Prestressed Concrete-Lined Pressure Tunnel

The purpose of prestressing works is to create a certain prestress in the final lining, so that it is free from tensile stresses induced by the internal water pressure during tunnel operation. The assessment of maximum internal water pressure depends on the prestress required for the given rock conditions, lining geometry and properties, creep, shrinkage and temperature changes at watering-up.

Principally, the grouting pressure applied into the circumferential gap, taking into account a certain factor of safety, must be maintained to a level below both the smallest principle stress in the rock mass and the compressive strength of concrete. Otherwise, the prestress induced in the final lining will be lost due to the opening of existing fissures in the rock mass.

After the tunnel excavation, consolidation grouting is a prerequisite that has to be accomplished prior to prestressing the final lining. This is preliminary meant to provide stability to the underground opening and reduce the permeability of the rock mass. As a result of consolidation grouting and prestressing works, the final lining, the shotcrete and the surrounding rock mass are in tight contact. Continuous load transfer from the final lining to the rock mass and vice versa can be preserved.

In 1985, Seeber introduced the load-line diagram method to determine the bearing capacity of prestressed concrete-lined pressure tunnels. In this method, the temporary support, such as shotcrete and anchors, should not permanently carry any loads.

In view of continuous contact between the final lining and the rock mass, the radial deformation of the final lining can be put equal to the radial deformation of the rock mass, referred to as the compatibility condition of deformations. Since the rock mass is assumed as an elastic isotropic material, the deformation in the lining is a function of tunnel geometry, elastic properties of the rock mass and of the lining. The load-line diagram method remains widely referred to and the most recent publication can be found in Wannenmacher et al. (2012).

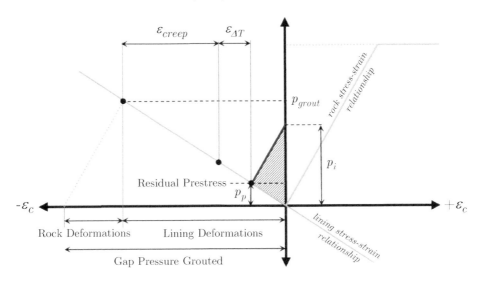

Fig. 3.2. The Modified Load-Line Diagram (after Seeber (1985a))

A concrete lining, which is not sealed with a plastic or waterproof membrane, is not absolutely impervious since radial cracks may develop during the hardening process of concrete. Concrete pores, together with radial cracks can permit seepage into the rock mass.

In contrast to impervious linings where the pressure head is dissipated at the lining intrados, pervious linings will allow seepage pressures that act not only in the lining but also in the rock mass. Seepage pressures can influence deformations in the rock mass, and therefore cannot be neglected when determining the bearing capacity of prestressed concrete-lined pressure tunnels(Schleiss, 1986b).

3.2. Prestress-Induced Hoop Strains

Since the grout is injected under high pressure, the circumferential gap between the final lining and the shotcrete is opened up and filled with densely compacted cement (Fig. 3.1). To obtain precise injections in the gap, circumferential and axial pipes are embedded along the tunnel walls. Debonding agents as well as synthetic membranes can be put on the shotcrete surface so as to ease the gap opening.

Because of consolidation grouting, the permeability of the rock mass will lower. With cement-based grout, the permeability of the grouted rock mass can be reduced to 10^{-7} m/s or about 1 Lugeon (Schleiss, 1986b). With stable grout (Fernandez, 1994) and micro-cements (Barton et al., 2001), the rock mass permeability close to 0.1 Lugeon can be achieved. Using artificial resins such as micro-silica and plasticizers in the grout mix, the permeability lower than 0.1 Lugeon is attainable (Barton, 2004), but this not economically attractive and the long-term behaviour of such artificial resins is not known.

In addition to reducing the permeability of the rock mass, consolidation grouting can potentially increase the modulus of elasticity of the rock mass. If well grouted, the modulus of elasticity of a fractured rock mass can at best be doubled (Jaeger, 1955; Kastner, 1962; Schwarz, 1985; Hendron et al., 1989). It has more effects on the loosened rock zone (Schleiss, 1986b; Schleiss and Manso, 2012) or at the location where low stresses are dominating (Barton et al., 2001; Vigl and Gerstner, 2009). However, if the modulus of elasticity of the rock mass is lower than that of the concrete, the modulus of elasticity of the grouted rock mass cannot be higher than that of the concrete (Schleiss, 1987; Hendron et al., 1989).

Based on the impervious thick-walled cylinder theory, the compressive hoop strain at the extrados of the final lining induced by the injection of grout at high pressure can be calculated as (Seeber, 1984; 1985a; 1985b; 1999):

$$\varepsilon_{\theta,\,p_{grout}}^{a} = \frac{p_{grout}}{E_c}\left(\frac{r_a^2 + r_i^2}{r_a^2 - r_i^2} - \nu_c\right) \tag{3.1}$$

where p_{grout} is the grouting pressure, r_i and r_a are the inner and outer radius of the final lining, E_c and ν_c are the modulus of elasticity and the Poisson's ratio of concrete, respectively.

Nowadays, the level of grouting pressure applied into the circumferential gap can be measured directly at the boreholes. Therefore, the pressure loss at the pump that was early introduced in the load-line diagram method can be omitted (Fig. 3.2). The highest strain loss in the final lining still remains due to shrinkage and creep. The creep in particular, can substantially relax the compressive stress that has already been induced in the final lining. Since shrinkage interacts with creep (Dezi et al., 1998), the total strain losses as a result of shrinkage and creep, ε_{creep}, can be taken between 30 and 40% (Seeber, 1985b).

The strain loss due to temperature changes at watering-up can be derived as a product of temperature change, ΔT, and the thermal coefficient for concrete, a_T. In the rock mass, approximately one-third of this loss can be expected. The strain losses as a result of temperature changes, $\varepsilon_{\Delta T}$, can be expressed as (Seeber, 1999):

$$\varepsilon_{\Delta T} = a_T \left(\Delta T + \frac{\Delta T}{3} \right) \tag{3.2}$$

Taking into account the compilation of strain losses, the effective prestress-induced hoop strain at the final lining extrados, $\varepsilon^a_{\vartheta, p_p}$, become:

$$\varepsilon^a_{\vartheta, p_p} = \varepsilon^a_{\vartheta, p_{grout}} - \varepsilon_{creep} - \varepsilon_{\Delta T} \tag{3.3}$$

Without internal pressure, the prestress-induced hoop strain in the final lining will reach its maximum value at the inner wall (Lauffer and Seeber, 1961; Timoshenko et al., 1970; Lu et al., 2011). The prestress-induced hoop strains at the final lining intrados, $\varepsilon^i_{\vartheta, p_p}$, can be obtained as:

$$\varepsilon^i_{\vartheta, p_p} = \varepsilon^a_{\vartheta, p_p} \left[\frac{2\, r_a^2}{(r_a^2 + r_i^2) - \nu_c\, (r_a^2 - r_i^2)} \right] \tag{3.4}$$

While the thickness of the shotcrete may vary between 5 and 10 cm, the thickness of the final lining should not be less than 25 cm to avoid radial cracks due to thermal cooling (Deere and Lombardi, 1989). In practice, the thickness of the final lining is between 30 and 35 cm. The minimum thickness of the final lining, $t_{c.\,min}$, with regard to the ultimate compressive strength of concrete, f_{cwk}, can be determined using:

$$t_{c.\,min} = r_a - \left[r_a^2 \left(1 - \frac{2\, p_{grout}}{0.75\, f_{cwk}} \right) \right]^{0.5} \tag{3.5}$$

3.3. Seepage-Induced Hoop Strains

If concrete is considered as a pervious material, its permeability can range from low to high depending on the care taken in the design and construction. According to Portland Cement Association (1979), the permeability of mature, good quality concrete without any minor cracks and construction irregularities is about 10^{-12} m/s. However, the permeability of an uncracked concrete lining without the implementation of waterstops at construction joints is normally from 10^{-7} to 10^{-8} m/s (Schleiss, 1997a).

Seepage per unit length, q, for a concrete-lined pressure tunnel situated above the groundwater level can be calculated iteratively using (Bouvard, 1975; Bouvard and Niquet, 1980; Schleiss, 1986b):

$$\frac{p_i}{\varrho_w \, g} - \frac{3}{4} r_g = \frac{q}{2\pi \, k_r} \ln \frac{q}{\pi \, k_r \, r_g} + \frac{q}{2\pi} \left[\frac{\ln \left(r_a / r_i \right)}{k_c} + \frac{\ln \left(r_g / r_a \right)}{k_g} \right] \quad (3.6)$$

where r_g is the radius of the loosened rock mass or the grouted zone, r_a is the outer radius of the final lining, r_i is the inner radius of the final lining, k_r is the permeability of the rock mass, k_g is the permeability of the grouted rock mass, k_c is the permeability of the concrete, p_i is the internal water pressure, g is the gravity acceleration and ϱ_w is the density of water.

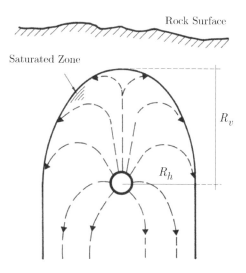

Fig. 3.3. Saturated Zone around a Pressure Tunnel (Schleiss, 1997b)

According to the continuity condition, seepage through an uncracked concrete lining, q_c, grouted zone, q_g, and rock mass, q_r, are equal, and can be calculated respectively using (Schleiss, 1986a):

$$q_c = \frac{(p_i - p_a) \, 2\pi \, k_c}{\varrho_w \, g \, \ln \left(r_a / r_i \right)} \quad (3.7)$$

$$q_g = \frac{(p_a - p_g) \, 2\pi \, k_g}{\varrho_w \, g \, \ln \left(r_g / r_a \right)} \quad (3.8)$$

$$q_r = \frac{(p_g - p_R) \, 2\pi \, k_r}{\varrho_w \, g \, \ln \left(R_v / r_g \right)} \quad (3.9)$$

in which p_a is the seepage pressure at the final lining extrados, p_g is the seepage pressure at the outer border of the grouted zone, and p_R is the seepage pressure in the rock mass influenced by the reach of the seepage flow.

Regarding seepage, its vertical, R_v, and the horizontal reach, R_h, can be estimated as (Schleiss, 1986b):

$$R_v = \left(\frac{q}{\pi \, k_r} \right) \ln (2)$$

$$R_h = \frac{q}{3 \, k_r} \tag{3.10}$$

The seepage pressure at the outer border of the grouted zone, p_g, can be calculated using (Bouvard, 1975; Simanjuntak et al., 2013):

$$\frac{p_g}{\varrho_w \, g} - \frac{3}{4} r_g = \frac{q}{2 \, \pi \, k_r} \ln \frac{q}{\pi \, k_r \, r_g} \tag{3.11}$$

Once the seepage pressure at the grouted zone is known, the seepage pressure at the final lining extrados, p_a, becomes:

$$p_a = p_g + \frac{q}{2 \, \pi \, k_g} \, \varrho_w \, g \ln (r_g / r_a) \tag{3.12}$$

Based on the porous thick-walled cylinder theory, the seepage-induced hoop strain at the final lining intrados, $\varepsilon^i_{\vartheta, \, p_i}$, can be calculated as (Schleiss, 1986b):

$$\varepsilon^i_{\vartheta, \, p_i} = \frac{(p_a - p_i)}{2 \, E_c \, (1 - \nu_c)} \left[\frac{1 + (r_a / r_i)^2}{(r_a / r_i)^2 - 1} + \frac{\ln (r_a / r_i) + (1 - 2 \, \nu_c)}{\ln (r_a / r_i)} \right] + \frac{2 \, p_F (r_a)}{E_c \, (1 - (r_i / r_a)^2)} \tag{3.13}$$

The mechanical boundary pressure at the final lining-grouted zone interface, $p_F(r_a)$, indicating the amount of pressure taken by the grouted zone, can be determined as (Schleiss, 1986b; Simanjuntak et al., 2012a):

$$p_F (r_a) = \frac{\begin{aligned} &\frac{(1 + \nu_g) \, E_c \, (p_g - p_a)}{(1 + \nu_c) \, E_g \, 2 \, (1 - \nu_g)} \cdot \\ &\cdot \left[\frac{r_a^2}{r_g^2 - r_a^2} \left(1 - 2 \, \nu_g + (r_g / r_a)^2 \right) + (1 - 2 \, \nu_g) \left(1 + \frac{1 - \nu_g}{\ln (r_g / r_a)} \right) \right] - \\ &- (p_a - p_i) \left[\frac{r_i^2}{r_a^2 - r_i^2} + \frac{(1 - 2 \, \nu_c)}{2 \ln (r_a / r_i)} \right] \end{aligned}}{\begin{aligned} &\frac{(1 + \nu_g) \, E_c}{(1 + \nu_g) \, E_c} \cdot \left[\frac{r_a^2}{r_g^2 - r_a^2} \left(1 - 2 \, \nu_g + (r_g / r_a)^2 \right) \right] + \\ &+ \left[\frac{2 \, r_i^2}{r_a^2 - r_i^2} \left(1 - \nu_c \right) + (1 - 2 \, \nu_c) \right] \end{aligned}} \tag{3.14}$$

in which E_g and ν_g denote the modulus of elasticity and the Poisson's ratio of the grouted rock mass, respectively.

3.4. Bearing Capacity of Prestressed Concrete-Lined Pressure Tunnels

Longitudinal cracks in the final lining can be avoided as long as the residual hoop strains at the final lining intrados during tunnel operation do not exceed the tensile strain of concrete. This criterion can be expressed as follows:

$$\varepsilon^i_{\vartheta,res} = \varepsilon^i_{\vartheta,p_p} + \varepsilon^i_{\vartheta,p,} < \frac{f_{ctk}}{E_c} \tag{3.15}$$

while $\varepsilon^i_{\vartheta,p_p}$ and $\varepsilon^i_{\vartheta,p_i}$ represent the prestress- and seepage-induced hoop strain at the final lining intrados respectively, f_{ctk} denotes the design tensile strength of concrete.

Considering that much of the tensile strength of concrete has already been used in the thermal cooling, Eq. (3.15) reduces to (Simanjuntak et al., 2012a):

$$\varepsilon^i_{\vartheta,p_p} + \varepsilon^i_{\vartheta,p_i} \leq 0 \tag{3.16}$$

Herein, the sign convention for compressive strains is negative.

3.5. Calculation Procedure

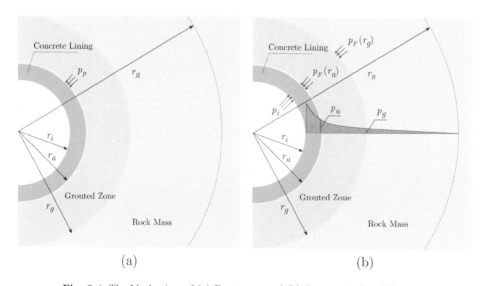

Fig. 3.4. The Mechanism of (a) Prestress-, and (b) Seepage- Induced Stresses

The calculation procedure to determine the bearing capacity of prestressed concrete-lined pressure tunnels as well as the seepage and seepage pressures around the tunnel is given as follows:

(A) As a result of prestressing works, calculate the prestress-induced hoop strain at the final lining intrados, $\varepsilon^i_{\theta,pp}$, using Eq. (3.4) by considering strain losses due to creep, shrinkage and temperature changes.

(B) Assume the internal water pressure, p_i, and calculate the seepage, q, using Eq. (3.6). At the same time, compute the seepage-induced hoop strain at the final lining intrados, $\varepsilon^i_{\theta,p_i}$, using Eq. (3.13) by taking into account the seepage pressure at the outer border of the grouted zone, p_g, obtained using Eq. (3.11) and seepage pressure at the final lining extrados, p_a, calculated using Eq. (3.12) as well as the mechanical boundary stress, $p_F(r_a)$, according to Eq. (3.14).

(C) Adjust the magnitude of the internal water pressure, p_i, established in step (B) until the criterion given by Eq. (3.16) is satisfied.

3.6. Practical Example

In the following, the proposed calculation procedure is implemented in an example. The main objectives are:

1. to determine the bearing capacity of a prestressed concrete-lined pressure tunnel
2. to quantify the seepage around the tunnel, and
3. to investigate the effects of the grouted zone on the stability of pressure tunnels

The pressure tunnel being considered has a circular geometry with an external radius of 2.30 m and is covered with an elastic isotropic rock mass subjected to a uniform in-situ stress of 40 MPa. The long-term stability of the pressure tunnel is ensured by using the passive prestressing technique, and consolidation grouting is executed up to a depth of 1 m behind the final lining.

The grouting pressure applied is 25 bar (2.5 MPa). While the losses due to shrinkage and creep are taken as 30%, the temperature change at watering-up is taken as 15° C. The tunnel is lined with concrete whose mechanical properties are according to type C25/30. Parameters used in the calculations are summarized in Table 3.1.

Table 3.1. Parameters Used in the Calculations

Material	Symbol	Value	Unit
	E_r	15	GPa
Rock Mass	ν_r	0.25	-
	k_r	10^{-6}	m/s
	E_q	15	GPa
Grouted Rock Mass	ν_q	0.25	-
	k_q	10^{-7}	m/s
	E_c	31	GPa
	ν_c	0.15	-
Concrete C25/30	f_{cwk}	30	MPa
(ÖNORM, 2001)	f_{ck}	22.5	MPa
	f_{ctm}	2.6	MPa
	f_{ctk}	1.8	MPa
	k_c	10^{-8}	m/s

3.6.1. Bearing Capacity of the Pressure Tunnel

When calculated using Eq. (3.5), the minimum thickness of the final lining is 27 cm. In the analysis, the lining thickness was taken as 30 cm in view of avoiding radial cracking during thermal cooling. Considering the strain losses due to shrinkage and creep, the effective grouting pressure acting on the lining was obtained as 7.1 bar (0.71 MPa). Based on Eq. (3.3), the prestress-induced hoop strain at the final lining extrados was calculated as 1.65×10^{-4}. By multiplying this value with a factor of 1.16 as presented in Eq. (3.4), the corresponding prestress-induced hoop strain at the intrados of the final lining, $\varepsilon^i_{\vartheta,\,p_p}$, became 1.91×10^{-4}. When assessed using the load-line diagram (Fig. 3.5), the maximum internal water pressure was obtained as 30.9 bar (3.09 MPa). This value needs to be evaluated in view of seepage effects on the rock mass.

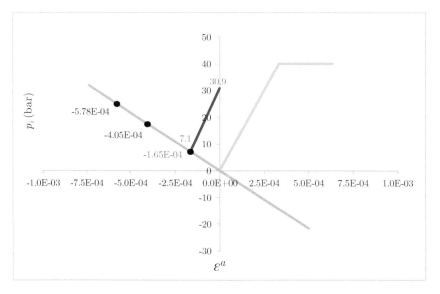

Fig. 3.5. The First Estimation of the Internal Water Pressure Calculated Using the Load-Line Diagram

Using Eq. (3.13), the corresponding seepage-induced hoop strain at the intrados of the final lining, $\varepsilon^i_{\vartheta,p_f}$, was calculated as 3.40×10^{-4}, ensuing the residual hoop strain in a tensile state of stress, $\varepsilon^i_{\vartheta,\,res}$, of $+1.48 \times 10^{-4}$. This strain corresponds to the tensile hoop stress of 4.5 MPa, which exceeds the tensile strength of concrete. Consequently, the internal water pressure has to be reduced in order to avoid longitudinal cracks in the lining.

To prevent the lining from continuous high levels of tensile stresses during operation and longitudinal cracks, the low tensile strength of concrete is not taken into account during the assessment of internal water pressure. Furthermore, construction joints may contain incipient longitudinal cracks, and much of the tensile strength of concrete has already been used during thermal cooling.

As the final lining has to be preserved in a compressive state of stress during tunnel operation, the internal water pressure obtained using the load-line diagram must be reduced. By putting the seepage-induced hoop strain at the final lining intrados equal to the prestress-induced hoop strain, the maximum internal water pressure, p_i, should not be greater than 17 bar (1.70 MPa) or the static water head should not exceed 173 m.

3.6.2. Seepage around Pressure Tunnel

According to Eq. (3.6), seepage, q, in the order of 55.70 l/s per km length of the tunnel occurs as a result of the 17-bar internal water pressure, which is still acceptable as long as the tunnel is not put at risks, such as in valley slopes. Due to seepage, a bell-shaped saturated zone can develop around the tunnel. Using Eq. (3.10), its vertical reach was calculated as 12.3 m, whereas its horizontal reach was 18.6 m measured from the tunnel centre.

Using Eqs. (3.7), (3.8) and (3.9), the seepage pressures around the pressure tunnel are obtained. The seepage pressure at the extrados of the final lining, p_a, was found as 28.50%p_i or equal to 4.84 bar (0.48 MPa) which is still below the smallest principal stress in the rock mass. The seepage pressure at the outer border of the grouted zone, p_g, was calculated as 10.03%p_i or equivalent to 1.70 bar (0.17 MPa), whereas in the rock mass it was 3.30%p_i or equal to 0.56 bar (0.06 MPa).

3.6.3. Effects of Grouted Zone on Stability of Pressure Tunnels

The purpose of consolidation grouting is to reduce the permeability of the rock mass and, if possible, to avoid disadvantageous alteration of the modulus of elasticity of the rock mass as a result of tunnel excavation. Besides resistance to leaching, the grout should be durable, has a sufficient compressive strength and low viscosity so as to penetrate fine joints. Consolidation grouting is prerequisite in view of limiting deformations of the loosened rock zone, as well as seepage.

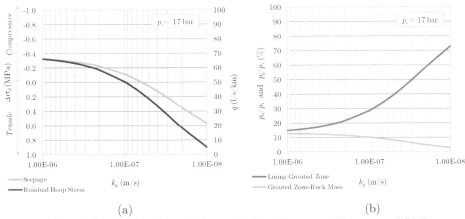

Fig. 3.6. Effect of Grouted Zone Permeability on (a) Hoop Stresses and Seepage, and (b) Seepage Pressures around the Tunnel

It has generally been acknowledged that consolidation grouting can reduce seepage. However, whether or not its efficacy can increase the rock strength is still arguable in view of limitations in grout properties and grouting technology (Barton et al., 2001). Therefore, it is important to note that mechanical effects of grouting are constrained and grouting can at best reinstate the modulus of elasticity of the loosened rock zone.

In the following, the influence of the grouted rock permeability, k_g, and the depth of grouting, r_g, on the bearing capacity of prestressed concrete-lined pressure tunnels is presented.

Table 3.2. Effects of Grouted Zone Permeability on Hoop Stresses, Seepage, Seepage Pressures and Seepage Reach

Output	k_g/k_c				
	100	50	10	5	1
$\Delta\sigma_\theta$ (MPa)	-0.32	-0.28	0.00	+0.23	+0.90
q (1/s/km)	66.28	64.92	55.70	47.25	21.13
p_a (bar)	2.54	2.83	4.84	6.69	12.39
p_g (bar)	2.16	2.10	1.70	1.36	0.48
R_v (m)	14.6	14.3	12.3	10.4	4.7
R_h (m)	22.1	21.6	18.6	15.8	7.0

Fig. 3.6 shows the effects of the permeability of the grouted zone on the residual hoop stresses at the final lining intrados and on the seepage around the tunnel. While grouting is favourable to reduce the amount of seepage around the pressure tunnel, it does not necessarily mean to increase the tunnel bearing capacity (Fig. 3.6a).

The less permeable the grouted zone, the higher the seepage pressure at the inner border of the grouted zone (Fig. 3.6b) and the greater the deformation will become at the intrados of the final lining. The results of residual hoop stress at the intrados of the final lining, seepage, seepage pressures, and seepage reach into the rock mass, for various permeability values of the grouted zone are summarized in Table 3.2.

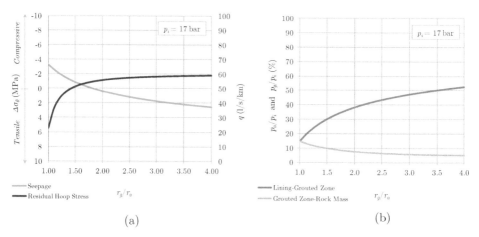

(a)   (b)

Fig. 3.7. Effect of Grouting Depth on (a) Hoop Stresses and Seepage, and (b) Seepage Pressures around the Tunnel

Analogously, by keeping the grouted zone permeability constant, the effects of the grouting depth on the bearing capacity of the pressure tunnel was investigated. The results are summarized in Table 3.3.

Table 3.3. Effects of Grouting Depth on Hoop Stresses, Seepage, Seepage Pressures and Seepage Reach

Output	r_g/r_a						
	1.0	1.5	2.0	2.5	3.0	3.5	4.0
$\Delta\sigma_\theta$ (MPa)	+5.44	-0.24	-1.14	-1.47	-1.64	-1.75	-1.80
q (l/s/km)	66.55	54.59	48.26	44.17	41.25	38.62	37.21
p_a (bar)	2.48	5.09	6.47	7.36	8.00	8.57	8.88
p_g (bar)	2.48	1.63	1.25	1.04	0.92	0.85	0.82
R_v (m)	14.7	12.0	10.7	9.8	9.1	8.5	8.2
R_h (m)	22.2	18.2	16.1	14.7	13.8	12.9	12.4

As illustrated in Fig. 3.7, the tensile hoop stress in the final lining and seepage can also be reduced by increasing the grouting depth. However, increasing the grouting depth beyond two times of the outer radius of the final lining is not practical and not worthwhile. Therefore, reducing the grouted zone permeability is more effective than increasing its depth. In practice, this can be achieved by shortening the distance among the boreholes.

Since the grout mix cannot penetrate discontinuities in the rock mass with a width greater than 0.1 mm (Schleiss, 1986b; Schleiss and Manso, 2012), it has to be noted that the permeability of the grouted zone lower than 10^{-7} m/s is difficult to achieve in practice. Unless artificial resins such as micro-silica and plasticizers are used in the grout mix, which will eventually lead to high construction costs.

3.7. Conclusions and Relevance

As long as the in-situ stresses in the rock mass are uniform and the assumption of elastic isotropic rock mass is acceptable, the load-line diagram method is useful for assessing the prestress-induced hoop strains in the final lining. However, since this method assumes impervious concrete lining and neglects seepage effects on the rock mass, it can result in overestimation of the maximum internal water pressure.

In this chapter, a new concept to determine the bearing capacity of prestressed concrete-lined pressure tunnels is introduced. The maximum internal water pressure was assessed by offsetting the seepage-induced hoop strain at the final lining intrados against the prestress-induced hoop strain. Once the maximum internal water pressure is obtained, a certain factor of safety has to be applied before putting the predicted value into practice.

Regarding concrete linings, they must have sufficient thickness and ability to deform. Whereas the former relates to anticipate radial cracks during thermal cooling, the latter concerns the outcome of prestressing works. Since concrete linings are pervious, seepage around pressure tunnels should always be expected. However, high seepage into the rock mass has to be prevented in view of tunnel safety.

Obviously, seepage into the rock mass depends on the permeability of the lining and the grouted zone. Therefore, consolidation grouting is a prerequisite when the passive prestressing technique is used to ensure the long-term stability of pressure tunnels. In view of its role particularly as a seepage barrier, improving grouting quality is more effective than increasing its depth.

The bearing capacity of pressure tunnels herein was determined based on the elastic response of the rock mass. Nevertheless, the rock mass failure can be controlled by numerous joint surfaces where plasticity cannot be overlooked. In such cases, a new approach considering the secondary stress field and effects of rock plasticity needs to be developed to more appropriately determine the bearing capacity of prestressed concrete-lined pressure tunnels.

4 | Pressure Tunnels in Uniform In-Situ Stress Conditions[2]

This chapter presents the numerical modelling of tunnel excavation, support installation, prestressing of final lining and activation of internal water pressure by means of a two-dimensional plane strain finite element model. The pressure tunnel stands in a deep elasto-plastic isotropic rock mass, whose in-situ stresses are uniform. While the Hoek-Brown failure criterion was employed to investigate the behaviour of the rock mass, the combined Rankine-Von Mises yield criteria was adopted to reveal stresses in the final lining.

Obviously, tunnel excavation is a three-dimensional problem. Herein, an approach is introduced so as to determine the load transmitted to the support when dealing with two-dimensional models. Since pervious concrete linings permit seepage into the rock mass, saturated zone develops around the tunnel. Furthermore, seepage pressures in the rock mass need to be assessed, in view of tunnel safety. For model validation, numerical results are compared with those obtained using the closed-form solutions.

[2] Based on Simanjuntak, T.D.Y.F., Marence, M., Schleiss, A.J., Mynett, A.E. (2012). *Design of Pressure Tunnels Using a Finite Element Model*. Hydropower & Dams 19(5): 98-105.

4.1. Introduction

As part of hydropower components, pressure tunnels play an important role in preserving the sustainability of hydropower operation. It is one of the most expensive constructions, especially when the traditional steel linings are used. The need for a more economical design of pressure tunnels has resulted in a shift from using steel to concrete linings. Nevertheless, the applicability of concrete-lined pressure tunnels is limited by the low tensile strength of concrete.

In cases of good quality of rock mass, the bearing capacity of concrete-lined pressure tunnels can be improved by injecting the circumferential gap between the final lining and the shotcrete with grout at high pressure, referred to as the passive prestressing technique. This is preliminary meant to provide enough compressive stresses in the final lining to suppress tensile stresses during tunnel operation. As well as economic benefits, the main advantage of this technique is that the final lining, the shotcrete and the surrounding rock mass are a composite construction (Fig. 4.1). Thereby, a continuous load transfer throughout the system can be provided.

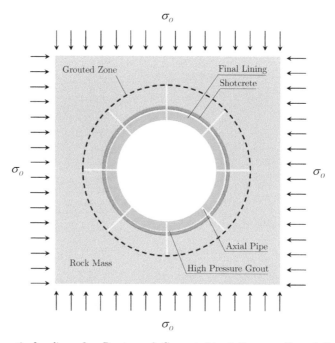

Fig. 4.1. Schematic Loading of a Prestressed Concrete-Lined Pressure Tunnel Embedded in an Elasto-Plastic Isotropic Rock Mass Subjected to Uniform In-Situ Stresses

Since it first appeared in Seeber (1985a; 1985b), the load-line diagram method has been used to determine the bearing capacity of prestressed concrete-lined pressure tunnels. This method remains widely referred to and the most recent publication can be found in Wannenmacher et al. (2012). This method is appropriate if the rock mass can be assumed as an elastic isotropic material. In reality, the rock mass failure can be controlled by numerous joint surfaces, where plasticity cannot be overlooked.

Because of the applicability of numerical models based on the finite element method in solving geotechnical problems dealing with non-linear deformability, material inhomogeneity and complex boundary conditions (Jing and Hudson, 2002; Jing, 2003), numerous finite element codes have nowadays been used to investigate the response of the rock mass to tunnelling. For cases of pressure tunnels, Stematiu et al. (1982) treated the rock mass as an elastic isotropic material, whereas Marence (1996) used the linear Mohr-Coulomb criterion with respect to an elasto-plastic framework. When using the Mohr-Coulomb criterion, the challenge remains to appropriately determine an equivalent angle of friction and cohesive strength for a given rock mass. Besides, a rock mass does not possess a linear behaviour (Hudson and Harrison, 2001). In this research, the non-linear Hoek-Brown criterion was employed so as to investigate the behaviour of the rock mass. The term non-linear failure criterion refers to the fact that the strength of the rock depends on the principal stresses in a non-linear manner (Clausen and Damkilde, 2008).

The objective of this research is to provide an overview of a practical application of a two-dimensional plane strain finite element model in investigating the behaviour of prestressed concrete-lined pressure tunnels. The pressure tunnel being considered is embedded in an elasto-plastic isotropic rock mass, whose in-situ stresses are uniform.

The numerical analyses are divided into four parts, consecutively dedicated to tunnel excavation, installation of support, lining prestressing, and activation of internal water pressure. Whereas the non-linear Hoek-Brown criterion is employed to investigate the behaviour of the rock mass, the combined Rankine-Von Mises yield criteria can be used to quantify prestress-induced hoop strains in the lining. For a proper estimation of the load transmitted to the support, a three-dimensional problem of tunnel excavation is considered in the model by using the convergence-confinement method. The internal water pressure is assessed using the superposition principle of strains at the final lining intrados. For model validation, the numerical results are compared with those obtained using the closed-form solutions.

4.2. The Hoek-Brown Failure Criterion

For cases when the assumption of elasto-plastic isotropic rock mass is reasonable, Hoek and Brown (1980a) suggest the expression regarding the strength of a rock mass as follows:

$$\sigma_1 = \sigma_3 + \sigma_{ci} \sqrt{m_b \frac{\sigma_3}{\sigma_1} + s} \qquad (4.1)$$

where σ_1 and σ_3 are the major and minor principal stress at failure respectively, σ_{ci} is the uniaxial compressive strength of the intact rock, m_b, and s are material constants that depend on the structure and surface conditions of the joints. The material constants m_b and s can be determined by using the following empirical formulae (Hoek et al., 2002):

$$m_b = m_i \, e^{(GSI - 100)/(28 - 14D)} \qquad (4.2)$$

$$s = e^{(GSI - 100)/(9 - 3D)} \qquad (4.3)$$

in which m_i is the petrographic constant, GSI is the Geological Strength Index and D is the blast damage factor. The value of GSI ranges from about 10 for extremely poor rock mass, to 100 for intact rock. The guidelines regarding GSI and D are presented in Fig. 4.2 and Table 4.1.

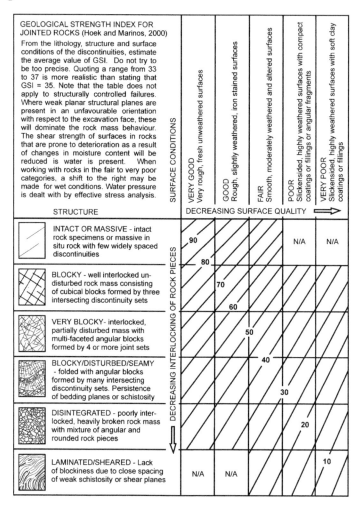

Fig. 4.2. General Chart for Estimating the GSI (Marinos et al., 2005)

The total deformations in the rock mass are made up of three components, namely elastic component, plastic component and volume increase in the plastic zone. The plastic potential is needed to control the volumetric change, in which its magnitude is characterized by a plastic dilation. The rate of plastic dilation is controlled by the parameter m_g, in which with relation to the dilation angle ψ, it can be calculated as (Clausen and Damkilde, 2008):

$$1 + m_g = \frac{1 + \sin \psi}{1 - \sin \psi} \tag{4.4}$$

Table 4.1. Suggested Value of D (Hoek et al., 2002)

Appearance of Rock Mass	Description	D
	Excellent quality controlled blasting or excavation by Tunnel Boring Machine (TBM) results in minimal disturbance to the confined rock mass surrounding the tunnel.	0
	Mechanical or hand excavation in poor quality rock masses (no blasting) results in minimal disturbance to the surrounding rock mass. Where squeezing problems result in significant floor heave, disturbance can be severe unless a temporary invert, as shown in the photograph, is placed.	0 0.5 no invert
	Very poor quality blasting in a hard rock tunnel results in severe local damage, extending 2 or 3 m, in the surrounding rock mass.	0.8
	Small scale blasting in civil engineering slopes results in modest rock damage, particularly if controlled blasting is used as shown on the left hand side of the photograph. However, stress relief results in some disturbance.	0.7 good blasting 1.0 poor blasting
	Very large open pit mine slopes suffer significant disturbance due to heavy production blasting and also due to stress relief from overburden removal. In some softer rocks excavation can be carried out by ripping and dozing, and the degree of damage to the slopes is less.	1.0 production blasting 0.7 mechanical excavation

In cases of plane strain in isotropic rocks with failure criteria independent of the main intermediate stress, the assumption of non-dilating rock mass, i.e. $\psi = 0°$, is appropriate (Wang, 1996; Hoek and Brown, 1997; Serrano et al., 2011), meaning that the rock mass undergoes no change in volume during plastic deformation. Another reason to consider this assumption is to avoid an overestimation of plastic dilation and thus plastic dissipation (Wan, 1992).

4.3. Excavation-Induced Stresses and Deformations

As a result of tunnel excavation, the principal stresses in the rock mass are disturbed and a new set of stresses are induced around the opening. The rock mass around the tunnel may not remain elastic anymore and can deform non-elastically.

To reveal excavation-induced stresses and deformations in an elasto-plastic isotropic rock mass, a two-dimensional finite element model can be used. For comparison, the closed-form solutions of excavation-induced stresses and deformations in an elasto-plastic isotropic rock mass are given herein.

4.3.1. Plastic Zone

The radius of elastic-plastic interface, R_{pl}, around a circular underground excavation, can be calculated using (Carranza-Torres and Fairhurst, 1999):

$$R_{pl} = R \, e^{\left[2\left(\sqrt{P_e^{cr}} - \sqrt{P_e}\right)\right]} \tag{4.5}$$

where P_e and P_e^{cr} denote the scaled tunnel support pressure and the scaled critical pressure, which can be obtained using the following equations respectively:

$$P_e = \frac{p_e}{m_b \, \sigma_{ci}} + \frac{s}{m_b^2} \tag{4.6}$$

$$P_e^{cr} = \frac{1}{16}\left[1 - \sqrt{1 + 16\left(\frac{\sigma_o}{m_b \, \sigma_{ci}} + \frac{s}{m_b^2}\right)}\right]^2 \tag{4.7}$$

If the critical pressure is higher than the scaled tunnel support pressure, plasticity will occur around the excavation.

4.3.2. Stresses and Deformations in the Elastic Region

The radial stresses at the elastic-plastic interface, $\sigma_{r,Rpl}$, can be calculated as:

$$\sigma_{r,R_{pl}} = \left(P_e^{cr} - \frac{s}{m_b^2}\right) m_b \, \sigma_{ci} \tag{4.8}$$

In the elastic region, the radial and hoop stresses as well as radial deformations given by the Lame's solution can be calculated respectively using the following equations (Sharan, 2005):

$$\sigma_r^{el} = \sigma_o - (\sigma_o - \sigma_{r,R_{pl}}) \left(\frac{R}{r}\right)^2 \tag{4.9}$$

$$\sigma_\vartheta^{el} = \sigma_o + (\sigma_o - \sigma_{r,R_{pl}}) \left(\frac{R}{r}\right)^2 \tag{4.10}$$

$$u_r^{el} = \frac{R^2}{r} \frac{(1+\nu)}{E_{rm}} (\sigma_o - \sigma_{r,R_{pl}}) \tag{4.11}$$

where σ_o represents the mean in-situ stress in the rock mass, ν is the Poisson's ratio of the rock, and E_{rm} denotes the rock mass modulus of deformation, which can be calculated using (Hoek et al., 2002):

$$E_{rm} = \left(1 - \frac{D}{2}\right) \sqrt{\frac{\sigma_{ci}}{100}} . 10^{[(GSI-10)/40]} \tag{4.12}$$

4.3.3. Stresses and Deformations in the Plastic Region

Based on the transformation rule for stresses and the assumption of ideally plastic behaviour of the rock mass, the radial and hoop stresses in the plastic region can be obtained respectively as (Carranza-Torres, 2004):

$$\sigma_r^{pl} = \left(\left[\sqrt{P_e^{cr}} + \frac{1}{2} \ln\left(\frac{r}{R_{pl}}\right)\right]^2 - \frac{s}{m_b^2}\right) m_b \, \sigma_{ci} \tag{4.13}$$

$$\sigma_\vartheta^{pl} = \left(\left[\sqrt{P_e^{cr}} + \frac{1}{2} \ln\left(\frac{r}{R_{pl}}\right)\right]^2 + \sqrt{P_e^{cr}} + \frac{1}{2} \ln\left(\frac{r}{R_{pl}}\right) - \frac{s}{m_b^2}\right) m_b \, \sigma_{ci} \tag{4.14}$$

Considering that there is no change in rock mass volume during plastic deformation, the radial deformations in the plastic zone can be determined using (Carranza-Torres and Fairhurst, 2000a):

$$\frac{u_r^{pl}}{R} \frac{E_{rm}}{(1+\nu)(\sigma_o - p_e^{cr})} = \left[\frac{1-2\nu}{2} \frac{\sqrt{P_e^{cr}}}{\left(\dfrac{\sigma_o}{m_b\,\sigma_{ci}} + \dfrac{s}{m_b^2}\right) - P_e^{cr}} + 1\right]\left(\frac{R_{pl}}{R}\right)^2 + \qquad (4.15)$$

$$+ \frac{1-2\nu}{4\left[\left(\dfrac{\sigma_o}{m_b\,\sigma_{ci}} + \dfrac{s}{m_b^2}\right) - P_e^{cr}\right]}\left[\ln\left(\frac{R_{pl}}{R}\right)\right]^2 -$$

$$- \frac{1-2\nu}{2} \frac{\sqrt{P_e^{cr}}}{\left(\dfrac{\sigma_o}{m_b\,\sigma_{ci}} + \dfrac{s}{m_b^2}\right) - P_e^{cr}}\left[2\ln\left(\frac{R_{pl}}{R}\right) + 1\right]$$

4.4. The Convergence Confinement Method

To provide stability to the underground opening, shotcrete can be sprayed onto the tunnel walls. Nowadays, shotcrete represents a principal support element of the New Austrian Tunnelling Method (Schütz et al., 2011; Schaedlich and Schweiger, 2014).

Principally, tunnel excavation is a three-dimensional problem. As a consequence, three-dimensional models are more appropriate to reveal stresses and displacements transmitted from the rock mass to the shotcrete. However, as long as one of the principal components of the in-situ stresses is acting parallel to the longitudinal axis of excavation, such problem can be solved using two-dimensional plane strain analysis taking into account stress relief occurring before shotcrete installation.

One of the methods to consider three dimensional tunnel advance and pre-relaxation ahead of the tunnel face in a plane strain analysis is the convergence-confinement method (Panet and Guenot, 1982). This method applies to symmetric problem of deep, uniformly supported, circular tunnels embedded in an isotropic rock mass subjected to uniform in-situ stresses. Because all field variables depend solely on the distance r measured from the tunnel axis, the problem becomes one-dimensional and therefore can be solved analytically (Cantieni and Anagnostou, 2011).

The three basic components of the convergence-confinement method (CCM) are: the ground reaction curve (GRC), which illustrates the relationship between the tunnel support pressure and the radial displacements at the wall, the support characteristic curve (SCC), which represents the stress-strain relationship in the support system, and the longitudinal deformation profile (LDP) which provides information regarding the stress relief occurring before support installation. Another benefit of the CCM is that the appropriate installation location of the support can be obtained with respect to elasto-plastic behaviour of the rock mass. The representation of the convergence-confinement method is presented in Fig. 4.3.

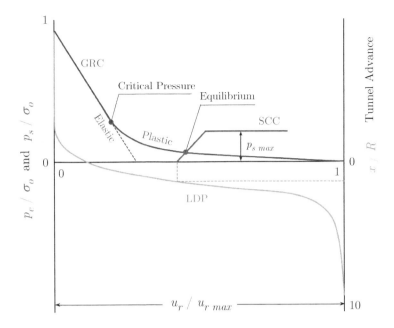

Fig. 4.3. Schematic Representation of the Convergence Confinement Method

While the GRC is obtained using Eqs. (4.5), (4.6), (4.7), (4.11) and (4.15), the SCC is the bearing capacity curve of the support. The support itself must have sufficient stiffness to maintain the excavation profile.

Due to its rapid hardening and ability to accept strains up to 1%, shorcrete is often used to support the excavation. For cases of hydropower tunnels, shotcrete with a thickness of 5 to 10 cm is common.

Assuming linear elastic material behaviour, the bearing capacity of shotcrete can be calculated using the relationship between the shotcrete stiffness and maximum sustainable stress (Hoek and Brown, 1980b). The elastic stiffness, K_s, and the maximum support pressure, $p_{s,max}$, of shotcrete can be calculated respectively using:

$$K_s = \frac{E_s}{(1 + \nu_s) R} \frac{R^2 - (R - t_s)^2}{(1 - 2\nu_s) R^2 + (R - t_s)^2} \tag{4.16}$$

$$p_{s,\max} = \frac{f_{ck}}{2} \left[1 - \frac{(R - t_s)}{R^2} \right] \tag{4.17}$$

The equilibrium condition (Fig. 4.3) is achieved when the retaining forces provided by the shotcrete is at least as great as the loading forces from the rock mass. The radial deformations of the shotcrete, u_s, can be expressed as:

$$u_s = \frac{p_s}{K_s} \tag{4.18}$$

To relate the tunnel deformations at the equilibrium stage to the actual physical location along the tunnel axis, the LDP is required. This is an important aspect of the design because the appropriate distance of shotcrete installation measured from the tunnel face can avoid errors resulting in shotcrete failure.

Based on a three-dimensional model, Vlachopoulos and Diederichs (2009) developed formulae for the LDP, taking into account the influence of plastic zones in the rock mass. In relation to the radial deformations, the points located ahead and behind the tunnel face can be obtained using Eqs. (4.19) and (4.20) respectively.

$$\frac{u_r}{u_{r,\max}} = \frac{e^{-0.15(R_{pl}/R)}}{3} \cdot e^{(x/R)} \tag{4.19}$$

$$\frac{u_r}{u_{r,\max}} = 1 - \left[\left(1 - \frac{e^{-0.15(R_{pl}/R)}}{3} \right) e^{-1.50(x/R_{pl})} \right] \tag{4.20}$$

Once the LDP is established, another question of maximum tolerable convergence arose since it is impossible in practice to install the shotcrete directly at the tunnel face. The tunnel wall convergence can be expressed in terms of a ratio between the radial deformation at the tunnel wall and the tunnel diameter. As suggested by Hoek (2000), the critical convergence of 1% should not be exceeded so as to avoid distress.

4.5. Bearing Capacity of Prestressed Concrete-Lined Pressure Tunnels

The long-term bearing capacity of concrete-lined pressure tunnels can be ensured by using the passive prestressing technique. For prestressed concrete-lined pressure tunnels, four zones can be considered (Fig. 4.1):

a. *rock mass*, which is undisturbed and can be assumed to behave as an elasto-plastic isotropic material.
b. *grouted zone*, is the area in the rock mass that needs to be grouted with cement-based grout. The grouted zone is economical if its external radius is between one and two times the tunnel radius. Although rock grouting is a common practice in tunnel construction, its effect is limited. Whereas the modulus of elasticity of a disturbed rock mass can at best be reinstated to that of an undisturbed by grouting, its permeability cannot be reduced below 10^{-7} m/s with cement-based grout.
c. *shotcrete*, is a special type of concrete and placed between the final lining and the grouted rock mass. It is responsible to support the rock mass after the excavation. Shotcrete is pervious and is often assumed as an elastic material. The modulus of elasticity of shotcrete is normally lower than that of concrete.

d. *final lining*, is made of concrete, which is pervious and in direct contact with the internal water pressure. To avoid tensile stresses during tunnel operation, the final lining is prestressed against the rock mass by injecting the circumferential gap between the final lining and the shotcrete with cement-based grout at high pressure.

The maximum internal water pressure can be assessed by offsetting the seepage-induced hoop strains at the final lining intrados against the prestress-induced hoop strains. In the final lining, the residual hoop strain should remain in a compressive state of stress during tunnel operation and can be expressed as:

$$\varepsilon_{\vartheta, p_p}^i + \varepsilon_{\vartheta, p_i}^i \leq 0 \tag{4.21}$$

4.6. Modelling of Pressure Tunnels

The modelling of pressure tunnels consists of tunnel excavation, shotcrete installation, prestressing of final lining, and activation of internal water pressure. In this research, the finite element code DIANA was used to reveal stresses and deformations in the rock mass as well as in the lining. Whereas the structural non-linear analysis implemented in DIANA was applied to reveal stresses and deformations, the steady-state groundwater analysis was used to predict seepage around the pressure tunnel.

As an example, consider an elasto-plastic isotropic rock mass, whose in-situ stresses are uniform. The excavation is assumed to result in minimal disturbance, or D equals to zero. The diameter of tunnel excavation is 4 m. The material properties for the rock mass are summarized in Table 4.2.

Table 4.2. Rock Mass Properties (Carranza-Torres and Fairhurst, 1999)

GSI	σ_{ci} (MPa)	m_i	m_b	s	ψ (°)	E_r (GPa)	ν_r
50	30	10	1.677	0.00387	0	5.5	0.25

The thickness of the shotcrete and the final lining are taken consecutively as 10 cm and 30 cm. Consolidation grouting is executed up to a depth of 1 m measured from the external radius of the shotcrete. Whereas the shotcrete has the material properties according to the concrete type C20/25, the final lining has the material properties corresponding to the concrete type C25/30. The relevant data for the shotcrete and the final lining are presented in Table 4.3.

Table 4.3. Concrete Properties (ÖNORM, 2001)

Type	E (GPa)	ν	f_{ctm} (MPa)	f_{ctk} (MPa)	f_{cuk} (MPa)	f_{ck} (MPa)
C20/25	20	0.15	2.2	1.5	25	18.8
C25/30	31	0.15	2.6	1.8	30	22.5

The tunnel being considered is subjected to uniform in-situ stresses of 20 MPa in the plane perpendicular to the tunnel axis. A two-dimensional plane strain is used, where model domain is made free to move in the radial direction, but not in the longitudinal direction.

4.7. Results and Discussions

4.7.1. Tunnel Excavation

In the model, the tunnel excavation can be simulated by deactivating the rock mass inside the tunnel. As a consequence, the initial stresses in the rock mass are changed and the first deformations occur around the underground opening.

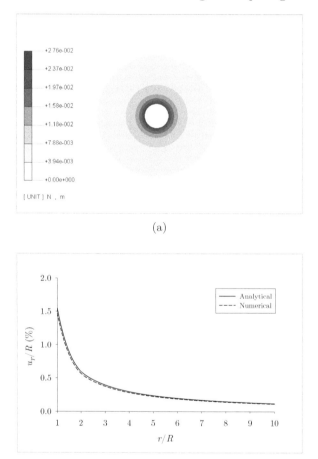

(a)

(b)

Fig. 4.4. Distribution of Excavation-Induced Radial Deformations

The numerical results of excavation-induced radial deformations are depicted in Fig. 4.4a. As great as 27.6 mm of deformations in a radial direction, which corresponds to the radial convergence $u_r/2R$ of 0.69%, was found at the tunnel walls. The comparison of results obtained using the numerical and analytical approach is presented in Fig. 4.4b. It can be seen that the numerical results show good agreement with those calculated using the closed-form solution.

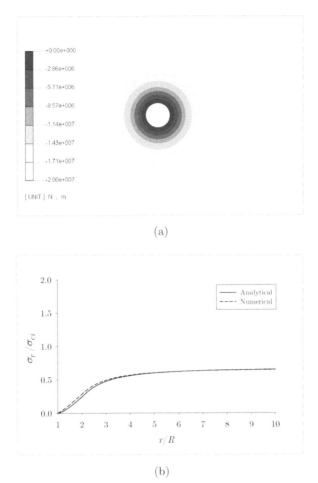

(a)

(b)

Fig. 4.5. Distribution of Excavation-Induced Radial Stresses

Fig. 4.5 illustrates the distribution of excavation-induced radial stresses around the tunnel. Herein, a compressive negative notation is used. As a result of tunnel excavation, the radial stresses in the rock mass decrease to zero in the direction towards the tunnel (Fig. 4.5a), meaning that the effects of excavation on the stresses in the rock mass decrease, as the distance r from the underground opening increases. The results of excavation-induced stresses obtained using the closed-form solution, are also given in Fig. 4.5b. The numerical model is seen to provide accurate results.

The distribution of excavation-induced hoop stresses in the space surrounding the rock mass is depicted in Fig. 4.6a. The classic jump representing the plastic-elastic interface in the rock mass was predicted to occur at $r/R = 2.2$ or at the distance of 4.4 m measured from the tunnel centre (Fig. 4.6b). The numerical results show that the hoop stress, $\sigma_{\theta i}$, at this interface was $1.00\sigma_{ci}$, whereas at the tunnel walls it was $0.16\sigma_{ci}$. It can also be seen that the numerical results fit the results calculated using the closed-form solution reasonably well.

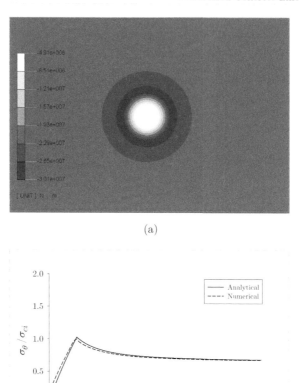

(a)

(b)

Fig. 4.6. Distribution of Excavation-Induced Hoop Stresses

4.7.2. Support Installation

To provide stability to the underground opening, a support system is needed. The load for which the support to be designed, is influenced by the three-dimensional tunnel advance and pre-relaxation ahead of the tunnel face. In this research, the load transmitted to the shotcrete taking into account the elasto-plastic behaviour of the rock mass as well as the appropriate location where the shotcrete has to be installed, was determined by means of the convergence-confinement method. Using the maximum critical convergence of 0.5% as the limit in order to avoid distress in the rock mass, the GRC, SCC and LDP are presented in Fig. 4.7.

The load transmitted to the 10-cm shotcrete was obtained at the point where the SCC intersects the GRC. As shown in Fig. 4.7, the pressure transmitted to the shotcrete was obtained as $0.036\sigma_o$ or 0.72 MPa. Furthermore, the shotcrete should be installed at a distance of $1.12R$ or 2.2 m measured from the tunnel face.

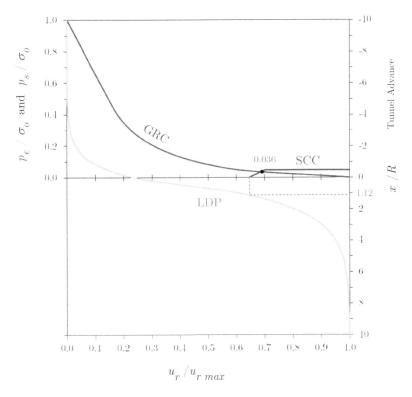

Fig. 4.7. Predicted Location of Shotcrete Installation

Once the pressure transmitted to the shotcrete has been determined by means of the convergence-confinement method, the shortcomings of two-dimensional models on predicting the deformations at the shotcrete-rock interface is solved. One of the modelling techniques is by reducing the modulus of elasticity of the rock mass inside the tunnel, referred to as the stiffness reduction method or α-method (Swoboda et al., 1993; Marence, 1996).

The modelling of support installation is independent from that of tunnel excavation. The rock mass inside the tunnel was activated but its stiffness was reduced such that the radial- deformations and stresses at the boundary between the shotcrete and the rock mass fit the results calculated using the convergence-confinement method.

The numerical results of radial deformations as a result of shotcrete installation are presented in Fig. 4.8a. When compared to the case of unsupported tunnel, the radial deformations at the tunnel walls were reduced to 20.9 mm or about 24%. This deformation corresponds to the radial convergence $u_r/2R$ of 0.5%, which is still below the maximum critical convergence according to Hoek (2000). As shown in Fig. 4.8b, the good agreement between the numerical and analytical results is evident. When the critical convergence is greater than 1%, the guidelines for the type and required amount of support systems suggested by Hoek and Marinos (2000) is recommended.

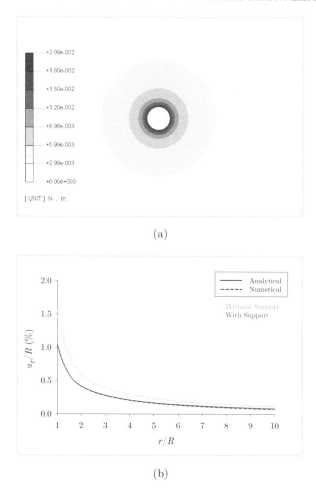

(a)

(b)

Fig. 4.8. Distribution of Radial Deformations in the Rock Mass after the Shotcrete Installation

The numerical results of radial stresses in the rock mass as a result of the installation of a 10-cm shotcrete are depicted in Fig. 4.9a. Since the tunnel is supported, as much as 0.72 MPa of pressure in a compressive state was found at the boundary between the shotcrete and the rock mass. The analytical results of radial stresses in the space surrounding the rock mass for the case of supported tunnel is presented in Fig. 4.9b. It can be seen that the numerical results are in good agreement with those calculated using the closed-form solution.

Similarly, the numerical results of hoop stresses in the rock mass due to the installation of a 10-cm shotcrete are illustrated in Fig. 4.10a. It is seen that as much as 8.46 MPa of hoop stresses in a compressive state was induced at the tunnel walls. The radius of the plastic-elastic interface in the rock mass measured from the tunnel centre was reduced from 4.4 m to 3.7 m or about 16% (Fig. 4.10b). The reduction of the plastic zone in the rock mass is favourable in view of tunnel stability.

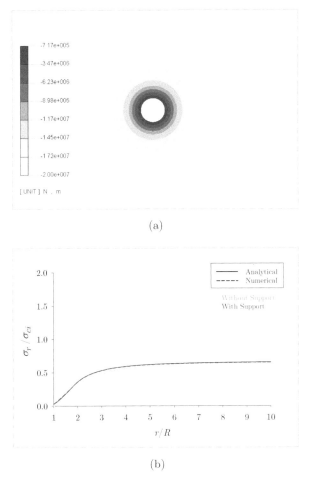

(a)

(b)

Fig. 4.9. Distribution of Radial Stresses in the Rock Mass after the Shotcrete Installation

According to the results presented in Figs. 4.7, 4.8, 4.9 and 4.10, it can be concluded that a 10-cm shotcrete installed at a distance of 2.2 m measured from the tunnel face can provide stability to the underground opening.

In view of equilibrium condition, the rock mass inside the tunnel is relaxed and the elements representing the shotcrete are activated. Thereby, the shotcrete takes the remaining load as a result of the tunnel excavation. Then, the final lining can be installed onto the shotcrete. To enhance the bearing capacity of the pressure tunnel, the final lining is prestressed using the passive prestressing technique. Since the final lining, the shotcrete and the rock mass are a composite construction due to grouting, the effective grouting pressure acting at the shotcrete-final lining interface can be assessed by using the load-line diagram method taking into account the strain losses due to creep, shrinkage and temperature changes.

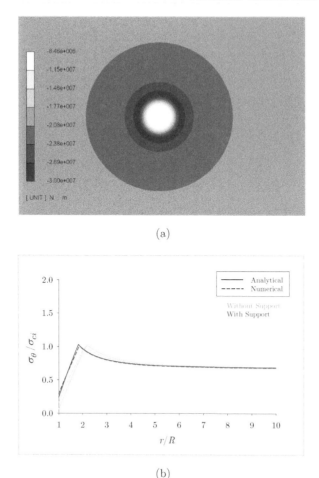

(a)

(b)

Fig. 4.10. Distribution of Hoop Stresses in the Rock Mass after the Shotcrete Installation

4.7.3. Prestressed Concrete Lining

Whereas the shotcrete absorbs part of the load as a result of the tunnel excavation, the final lining, together with the rock mass, withstands the internal water pressure. A 30-cm final lining can be installed onto the shotcrete by beforehand completing the consolidation grouting. This is preliminary meant to induce favourable changes of the rock mass properties around the tunnel.

In practice, the lining prestressing can be executed in form of injections around the tunnel. By pumping the grout at high pressure into the contact face between the shotcrete and the final lining, the circumferential gap between the final lining and the shotcrete is opened up and filled with densely compacted cement. As a result, the final lining is prestressed against the shotcrete and the rock mass and at the same time a full contact in the system is provided as the grout hardens.

The prestressing of the final lining is modelled according to the concept of compatibility conditions provided that the shotcrete and the final lining are continuous with no slip conditions at their interface. The grouting pressure acting at the extrados of the final lining can be determined by using the load-line diagram taking into account pressure losses due to creep, shrinkage and temperature effects.

In this research, the gap grouting was modelled by applying a uniform compressive pressure along the shotcrete-final lining interface. In order to avoid the final lining is affected by the previous deformations, the radial deformations at the shotcrete-final lining interface must be reset to zero before applying the grouting pressure. To reveal stresses in the shotcrete and the final lining as a result of gap grouting, the combined Rankine-Von Mises yield criteria (Feenstra, 1993) was used. Whereas the former criterion describes the tensile regime, the latter controls the compressive regime.

In practice, the final lining is prestressed after the completion of consolidation grouting. It is considered that some quite moderate improvements occur to the permeability of the surrounding rock mass due to consolidation grouting. These improvements were assumed to take effect up to a depth of 1 m behind the shotcrete. The permeability coefficients for the final lining, k_c, the shotcrete, k_s, the grouted rock mass, k_g, and the ungrouted rock mass, k_r, are listed in Table 4.4.

Table 4.4. Permeability Coefficients Used in the Analysis

k_c (m/s)	k_s (m/s)	k_g (m/s)	k_r (m/s)
10^{-8}	10^{-8}	10^{-7}	10^{-6}

With regard to the smallest in-situ stress in the rock mass, creep, shrinkage and temperature effects, the effective grouting pressure, p_p, acting at the shotcrete-final lining interface was assumed as 10 bar (1 MPa). The numerical results of prestress-induced radial stresses and hoop stresses in the final lining are shown in Fig. 4.11a and b, respectively.

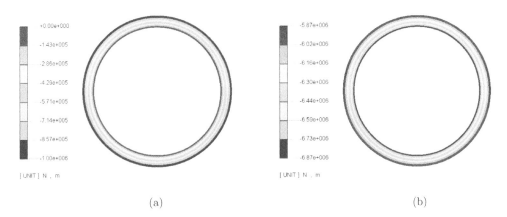

(a)   (b)

Fig. 4.11. Distribution of (a) Prestress-Induced Radial Stresses, and (b) Prestress-Induced Hoop Stresses in the Final Lining

As a result of prestressing works, as much as 1 MPa of compressive stress in a radial direction was active along the shotcrete-final lining interface. In view of uniform in-situ stresses in the rock mass, the distribution of radial stresses in the final lining is uniform. At the perimeter of the final lining intrados, the prestress-induced radial stress is zero (Fig. 4.11a).

On the contrary, the prestress-induced hoop stress is minimum at the final lining extrados and is maximum at the final lining intrados. While the prestress-induced hoop stress at the final lining extrados was obtained as 5.87 MPa, it was 6.87 MPa at the final lining intrados (Fig. 4.11b). If calculated using the thick-walled cylinder theory (Timoshenko et al., 1970), comparable results are obtained. The results are summarized in Table 4.5.

Table 4.5. Stresses in the Final Lining

Method	σ_r (MPa)		σ_θ (MPa)	
	$r = r_i$	$r = r_a$	$r = r_i$	$r = r_a$
FEM	0	1.00	6.87	5.87
Thick-Walled Cylinder	0	1.00	6.88	5.88

4.7.4. Activation of Internal Water Pressure

The maximum internal water pressure can be assessed using the superposition principle of hoop strains given by Eq. (4.21). The residual hoop strains at the final lining intrados should be kept not to fall in a tensile state of stress during tunnel operation in order to avoid lining cracking. Moreover, minor cracks can occur in the rock mass, especially in the plastic zone. Therefore, the rock mass cannot take the hoop stresses from the final lining, but only the radial stresses.

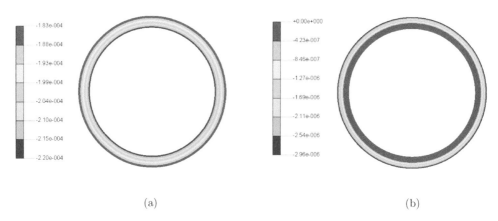

(a) (b)

Fig. 4.12. Distribution of Hoop Stresses in the Final Lining (a) Before, and (b) After the Loading of Internal Pressure

To simulate seepage around the pressure tunnel, two permeable boundaries located at the final lining intrados and at the outer border of the model domain were set up in the model. While the former represents the hydrostatic head due to the internal water pressure, the latter relates to the hydrostatic head imposed by the groundwater level. In this research, the tunnel is situated above the groundwater level.

To assess the maximum internal water pressure, the hydrostatic head inside the pressure tunnel were increased to a level where the seepage-induced hoop strains at the final lining intrados offset the prestress-induced hoop strains. As soon as the residual hoop strain at the final lining intrados is zero, the maximum internal water pressure is reached.

As shown in Fig. 4.12a, as much as 0.22‰ of compressive hoop strains was active at the final lining intrados before the activation of internal water pressure. This value decreased to zero when as high as 117 m of hydrostatic head that is equivalent to the internal water pressure, p_i, of 1.15 MPa, was activated inside the tunnel. At the same time, the compressive strains at the final lining extrados decreased from 0.18‰ to nearly zero, i.e. 2.96×10^{-6} (Fig. 4.12b). However, these strains still remained in a compressive state of stress.

In view of permeable concrete linings, seepage will occur around the tunnel. The rate of seepage depends not only on the internal water pressure, but also on the permeability of the final lining, the shotcrete, the grouted zone, and the rock mass.

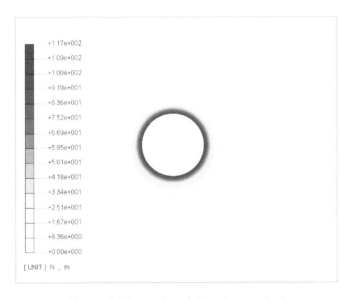

Fig. 4.13. Distribution of Pore Pressure Head

The numerical result illustrating the distribution of pore pressure head around the pressure tunnel is presented in Fig. 4.13. The pressure head at the extrados of the final lining, the shotcrete as well as the grouted zone was obtained as 48.2 m, 27.6 m, and 11.3 m, respectively, which corresponds to the seepage pressure of 4.73 bar (0.47 MPa), 2.71 bar (0.27 MPa), and 1.11 bar (0.11 MPa).

These seepage pressures result in seepage, q, of 25.24 l/s per km length of the tunnel, which is equal to 2.19 l/s/km/bar. As comparison, seepage, q, in the order of 26.37 l/s per km length of the tunnel, which corresponds to 2.29 l/s/km/bar was obtained when calculated using the formulae presented in Schleiss (1986b). This seepage is still tolerable as long as hydraulic jacking or fracturing of the rock mass can be avoided and the pressure tunnel is not put at risks, such as being constructed in valley slopes or in a location that can negatively influence the hydrogeological conditions.

4.8. Conclusions

The design of prestressed concrete-lined pressure tunnels features delicate phenomena of not only the rock mass, but also the support and the final lining. Whereas two-dimensional models are capable in simulating the tunnelling construction process and thus are often used, some of the limitations of two-dimensional models relate to the proper determination of the load transferred to the support.

By means of the convergence-confinement method, an attempt has been made herein to solve such problem. This research suggests that this method can be incorporated in two-dimensional models to determine the part of load transferred to the support and the appropriate location of the support installation with respect to elasto-plastic behaviour of the rock mass. This is an important aspect in the design of prestressed concrete-lined pressure tunnels since a plastic zone containing minor cracks cannot take tensile hoop stresses from the lining. However, it has to be acknowledged that the use of the convergence-confinement method is restricted to cases of circular tunnels embedded in an isotropic rock mass, whose in-situ stresses are uniform.

This chapter illustrates the applicability of a two-dimensional finite element model to assess the bearing capacity of prestressed concrete-lined pressure tunnels. Based on the superposition principle of hoops trains, that is the sum of prestress- and seepage-induced hoop strains at the final lining intrados, the maximum internal water pressure can be assessed. However, in view of uncertainties, a certain factor of safety has to be applied to the predicted value before putting it into practice. In view of pervious linings, the predicted seepage and saturated zone around the tunnel can also be obtained.

The benefits of the combination of analytical and numerical approach have been presented in this chapter. In overall, there is a global coherence between the results calculated using the closed-form solution and those obtained using the numerical model. On the one hand this implies that two-dimensional models are potential to reproduce tunnel behaviour in a realistic manner, on the other hand this shows that analytical solutions, regardless their simplifications, possess great value for conceptual understanding of tunnelling construction processes and for model validation, and therefore cannot be overlooked.

5 | Pressure Tunnels in Non-Uniform In-Situ Stress Conditions[3]

The bearing capacity of prestressed concrete-lined pressure tunnels is governed by the in-situ stresses in the rock mass, which generally have different magnitudes in the vertical and horizontal direction. Two cases were distinguished based on whether the in-situ vertical stress is greater than the in-situ horizontal stress, or not.

The distribution of stresses and deformations in the rock mass and the lining due to the tunnelling construction process was investigated by means of a two-dimensional finite element model. The effects of the in-situ stress ratio on the bearing capacity of prestressed concrete-lined pressure tunnels embedded in an elasto-plastic isotropic rock mass were explored. Favourite locations where longitudinal cracks may occur in the final lining were identified, which is useful for taking measures regarding tunnel tightness and stability.

---

[3] Based on Simanjuntak, T.D.Y.F., Marence, M., Mynett, A.E., Schleiss, A.J. (2014). *Pressure Tunnels in Non-Uniform In-Situ Stress Conditions.* Tunnelling and Underground Space Technology 42, 227-236.

5.1. Introduction

The in-situ stress in a rock mass is one of the most important aspects in the design of deep tunnels. Its magnitudes, which are generally unequal in the vertical and horizontal direction, can influence the response of the rock mass to excavation and thus the bearing capacity of prestressed concrete-lined pressure tunnels.

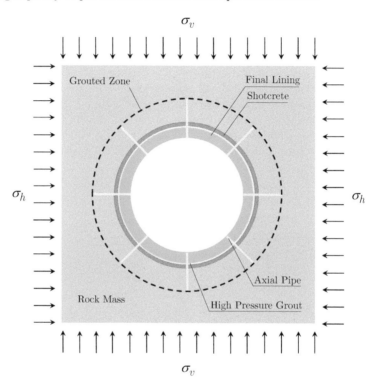

Fig. 5.1. Schematic Loading of Prestressed Concrete-Lined Pressure Tunnel Embedded in an Elasto-Plastic Isotropic Rock Mass Subjected to Non-Uniform In-Situ Stresses

As long as the in-situ stresses normal to the axis of a circular tunnel are uniform, the convergence-confinement method can be applied to assess excavation-induced stresses and deformations transferred to the support. Carranza-Torres and Fairhurst (2000a) summarized the three basic components of the convergence-confinement method, i.e. the ground reaction curve (GRC), the support characteristic curve (SCC), and the longitudinal deformation profile (LDP) for the design of support for tunnels embedded in an elasto-plastic isotropic rock mass.

Concurrently, Simanjuntak et al. (2012b) have demonstrated the applicability of a two-dimensional plane strain finite element model to determine the bearing capacity of prestressed concrete-lined pressure tunnels subjected to uniform in-situ stresses. Their study involved the convergence-conferment method and showed that there is a global coherence of numerical results when compared to those calculated by using the closed-form solutions.

Analytical approaches solving cases of prestressed concrete-lined pressure tunnels embedded in an elasto-plastic isotropic rock mass subjected to non-uniform in-situ stresses (Fig. 5.1) are still lacking and not explicitly revealed. The implementation of numerical models can be effective in gaining a better understanding of tunnel behaviour and may promise benefits in many areas including hydropower tunnels.

This chapter is aimed at investigating the mechanical and hydraulic behaviour of prestressed concrete-lined pressure tunnels subjected to non-uniform in-situ stresses of the rock mass based on numerical analyses. To be concordant with the previous work, this research is dedicated to a deep, straight ahead circular tunnel situated above the groundwater level.

The rock mass supporting the tunnel is isotropic and assumed to behave as an elasto-plastic non-dilatant material. After the excavation-induced stresses and deformations have been obtained, the analysis progresses to the assessment of stresses and deformations transmitted to the shotcrete. The final lining is concreted onto the shotcrete and is prestressed by using the passive prestressing technique. The maximum internal water pressure is assessed based on the superposition principle of hoop strains in the final lining. Due to pervious concrete, seepage occurs around the pressure tunnel and it needs to be quantified in view of tunnel safety.

5.2. Non-Uniform In-Situ Stresses in the Rock Mass

In cases where there is no preferred orientation of joints within the rock mass so that the assumption of elasto-plastic isotropic rock mass is acceptable, the Hoek-Brown failure criterion is appropriate. The strength of a rock mass according to the Hoek-Brown failure criterion is expressed as (Hoek and Brown, 1980b):

$$\sigma_1 = \sigma_3 + \sigma_{ci} \sqrt{m_b \frac{\sigma_3}{\sigma_1} + s} \tag{5.1}$$

in which σ_{ci} is the uniaxial compressive strength of the intact rock material, σ_1 and σ_3 represent the major and minor principal stress respectively. Parameter constants m_b and s depend on the structure and surface conditions of the joints and can be evaluated using the Geological Strength Index (GSI).

For cases of deep tunnels, the variation of vertical loading across the height of excavation is small compared to the magnitude of stresses at the excavation location, and therefore, the gravitational force is negligible (Detournay and Fairhurst, 1987). The in-situ horizontal stress, σ_h, can be expressed in the product of the corresponding in-situ vertical stress, σ_v, and a coefficient of earth pressure at rest, k.

The in-situ stresses in the rock mass are non-uniform, if $k \neq 1$. In the plane strain conditions, the mean in-situ stress, σ_o, is calculated by using:

$$\sigma_o = \frac{\sigma_h + \sigma_v}{2} = \frac{k \sigma_v + \sigma_v}{2} = \frac{(k + 1) \sigma_v}{2} \tag{5.2}$$

As a result of tunnel excavation, a plastic zone can develop around a tunnel. Its volumetric behaviour is characterized by a dilation angle, ψ. For cases of plane strain in isotropic rocks, the assumption of non-dilating rock mass, i.e. $\psi = 0$, is appropriate for the prediction of plastic zone (Wang, 1996; Hoek and Brown, 1997; Serrano et al., 2011).

5.3. Tunnel Excavation in Elasto-Plastic Rocks

In this chaper, there are two cases considered: a case where the in-situ vertical stress is greater than the horizontal and another case, where the in-situ horizontal stress is greater than the vertical. For both cases, the mechanical properties of the rock mass adopted from Amberg (1997) are listed in Table 5.1. The radius of tunnel excavation, R, is 2 m.

Table 5.1. Rock Mass Properties (Amberg, 1997)

GSI	σ_{ci} (MPa)	m_i	m_b	s	ψ (°)	E_r (GPa)	ν_r
65	75	17	4.87	0.02	0	20.5	0.25

The state of stress in the rock mass is characterized by a compressive mean stress, σ_o, equals to 40 MPa. The rock mass parameters m_b, s and E_r can be obtained using the program RocLab (2002) according to the formulae given by Hoek et al. (2002). The finite element code DIANA was used in this research and the structural non-linear analysis was selected to simulate the tunnel excavation focussing on the influence of non-uniform in-situ stresses and the effects of elasto-plastic rock mass yield on the stresses and deformations distribution. A two-dimensional plain strain condition was assumed meaning that the out-of-plane stress coincides with the intermediate principal stress σ_2, and that the problem geometry being analysed is long and has a regular cross-section in the out-of-plane direction (Eberhardt, 2001). The sign convention for compressive stresses is negative.

Before the rock mass is excavated, the in-situ stresses are undisturbed. After the removal of the rock mass, the first deformations occur around the tunnel. Because of a general property of elasto-plastic continua, according to which the displacements depend linearly on $1/E$ (Anagnostou and Kovari, 1993; Schürch and Anagnostou, 2012), the excavation-induced radial deformations can be expressed as:

$$\frac{E_r \, u_r}{\sigma_o \, R} = f\left(\frac{\sigma_{ci}}{\sigma_o}, k, \nu, m_b, s, \psi\right) \tag{5.3}$$

The numerical results of excavation-induced radial deformations are shown in Fig. 5.2. While Fig. 5.2a illustrates the results of radial deformations after the tunnel excavation for the case where the horizontal-to-vertical stress coefficient $k = 0.80$, Fig. 5.2b shows the results for $k = 1.25$. For both cases, the results of radial deformations along the tunnel perimeter in a polar system of coordinates are presented in Fig. 5.3. Comparing the results between Fig. 5.3a and b, unlike the results for cases where the in-situ stresses are uniform ($k = 1$), the radial deformations for cases where the in-situ stresses are non-uniform, are non-uniformly distributed.

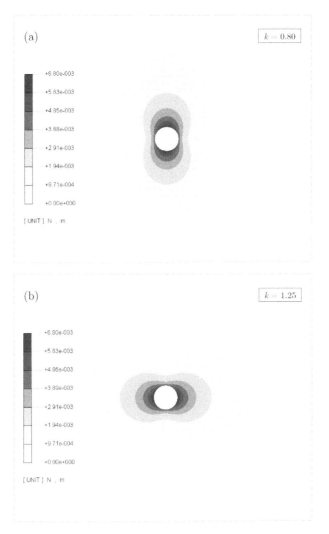

Fig. 5.2. Radial Deformations after the Tunnel Excavation

The shape of the underground opening for cases where the in-situ stresses are non-uniform is different from those for cases where the in-situ stresses are uniform. The results in Fig. 5.3a suggest that, if the in-situ vertical stress is greater than the in-situ horizontal stress, the radial deformations at the roof and invert are greater than those at the sidewalls. This indicates that the shape of the tunnel is oval with its major axis perpendicular to the direction of in-situ vertical stress. Comparable tunnel behaviour was observed by González-Nicieza et al. (2008) when investigating the influence of depth and shape of tunnels on the distribution of radial deformations. In a similar way, if the in-situ horizontal stress is greater than the in-situ vertical stress, the radial deformations at the sidewalls are greater than those at the roof and invert (Fig. 5.3b). Here, the shape of the tunnel is still oval, but with its major axis parallel to the direction of in-situ vertical stress.

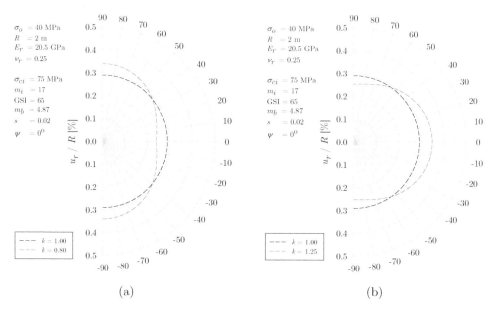

Fig. 5.3. Distributions of Excavation-Induced Radial Deformations along the Tunnel Perimeter

If $k < 1$, the maximum radial deformation is located at the tunnel roof and invert, while the minimum radial deformation is found at the sidewalls. For a ratio $k = 0.80$ (Fig. 5.3a), the corresponding radial convergence at the tunnel roof and invert was obtained as $u_r/2R = 0.17\%$, while at the sidewalls it was 0.12%. Conversely, if $k > 1$, the maximum radial deformation is situated at the tunnel sidewalls, whereas the minimum radial deformation is located at the roof and invert. The radial convergence at the sidewalls for $k = 1.25$ (Fig. 5.3b) was found as 0.17%, while at the roof and invert it was 0.12%.

The numerical results of excavation-induced radial stresses around the tunnel for both $k = 0.80$ and 1.25 are depicted in Fig. 5.4a and b respectively. It is seen that the change of stress level in the rock mass is more profound around the sidewalls, when the in-situ vertical stress is greater than the in-situ horizontal stress. When the in-situ horizontal stress is greater than the in-situ vertical, the change of stress level is more profound around the roof and invert. However, the discrepancy of the change of stress level decreases towards the underground opening. Since the tunnel is unsupported, radial stresses at the tunnel walls diminish to zero.

As shown in Figs. 5.2 and 5.4, the distribution of excavation-induced radial deformations and radial stresses for a specific value of the horizontal-to-vertical stress coefficient, k, is similar to the case with coefficient $1/k$ by rotating the tunnel axis by 90°. Here, the response of the rock mass to circular excavation is comparable to that has been reported in Carranza-Torres and Fairhurst (2000a).

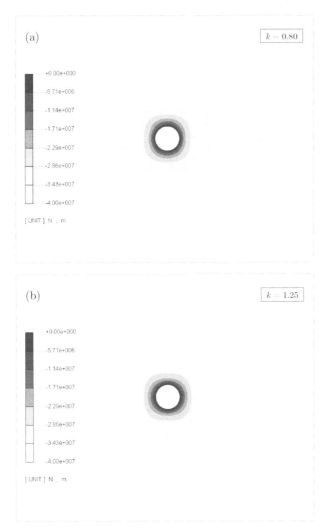

Fig. 5.4. Radial Stresses after the Tunnel Excavation

5.4. Radial Stresses and Deformations Transmitted to a Support System

The radial deformations in the rock mass due to tunnel excavation can be limited by applying a support system. If the critical convergence is less than 1%, a 5 to 10 cm shotcrete is commonly installed to support the underground opening for hydropower tunnels. For tunnels whose the critical convergence is greater than 1%, the guidelines for the type and required amount of support is available in Hoek and Marinos (2000).

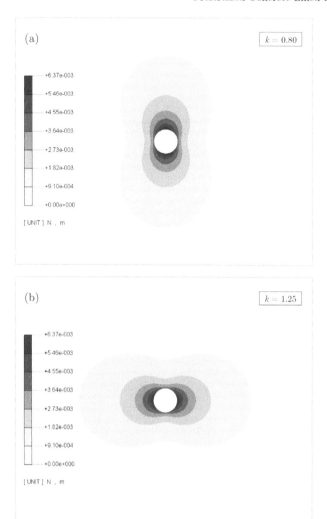

Fig. 5.5. Radial Deformations after the Shotcrete Installation

In addition to the rock mass parameters, the magnitude of stresses transmitted by the rock mass to the shotcrete is governed not only by the modulus of elasticity, E_s, the compressive strength, f_{ck}, and the thickness of the shotcrete, t_s, but also by the distance, x, from the tunnel face at which the shotcrete is installed. The radial deformations at the rock-shotcrete interface can be expressed as:

$$\frac{E_r\, u_r}{\sigma_o\, R} = f\left(\frac{\sigma_{ci}}{\sigma_o}, \frac{f_{ck}}{E_s}, \frac{x}{t_s}, k, \nu, m_b, s, \psi\right) \tag{5.4}$$

Table 5.2. Concrete Properties (ÖNORM, 2001)

Type	E (GPa)	ν	f_{ctm} (MPa)	f_{ctk} (MPa)	f_{cwk} (MPa)	f_{ck} (MPa)
C20/25	20	0.15	2.2	1.5	25	18.8
C25/30	31	0.15	2.6	1.8	30	22.5

As an example, the concrete properties of C20/25 listed in Table 5.2 according to ÖNORM (2001) can be adopted for shotcrete. The distributions of radial deformations and radial stresses around the tunnel after the installation of a 10 cm shotcrete for both $k = 0.80$ and $k = 1.25$, are presented in Figs. 5.5 and 5.7 respectively.

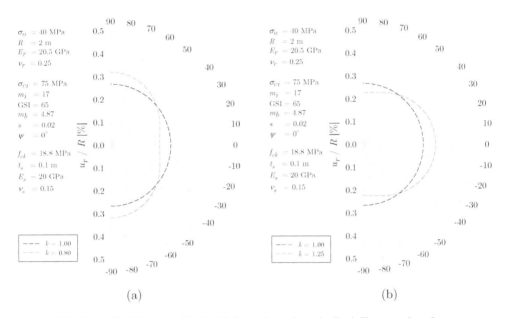

(a)   (b)

Fig. 5.6. Distributions of Radial Deformations along the Rock-Shotcrete Interface

Fig. 5.6 shows the distributions of radial deformations along the rock-shotcrete interface after the shotcrete installation. For $k = 0.80$ (Fig. 5.6a), the radial convergence at the tunnel roof and invert was reduced to 0.16%, while at the sidewalls it was 0.11%. For $k = 1.25$, reverse values were obtained. The radial convergence at the tunnel roof and invert was decreased to 0.11%, whereas at the sidewalls it was 0.16% (Fig. 5.6b).

The distributions of radial stresses along the rock-shotcrete interface for $k = 0.80$ and $k = 1.25$ are depicted respectively in Fig. 5.8a and b. When $k < 1$, the radial stresses at the sidewalls are greater than those at the roof and invert, whereas if $k > 1$, the radial stresses at the roof and invert are greater than those at the sidewalls.

In the following, the numerical results of radial stresses for both $k = 0.80$ and 1.25 are analysed. When $k = 0.80$ (Fig. 5.8a), the radial stress, σ_r/σ_{ci}, transmitted to the shotcrete at the roof and invert was found as 1.4%, whereas at the sidewalls it was

1.5%. When $k = 1.25$ (Fig. 5.8b), reverse values from those obtained for $k = 0.80$ were obtained. Again, by rotating the tunnel axis by 90°, the magnitude of radial stresses transmitted to the shotcrete for a specific value of k is similar to that of $1/k$.

The numerical results presented in Figs. 5.6 and 5.8 were obtained by adopting the same setting as proposed by Carranza-Torres and Fairhurst (2000b), assuming that the shotcrete is installed in continuous contact with the surrounding rock mass before the in-situ stresses are relaxed. Since the stress relief occurs ahead the excavation is not considered, it has to be acknowledged that this approach may lead to an overestimation of stresses transmitted to the shotcrete. Nevertheless, it would be beneficial in view of a safer shotcrete design.

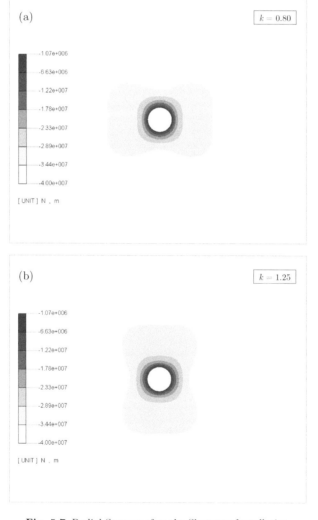

Fig. 5.7. Radial Stresses after the Shotcrete Installation

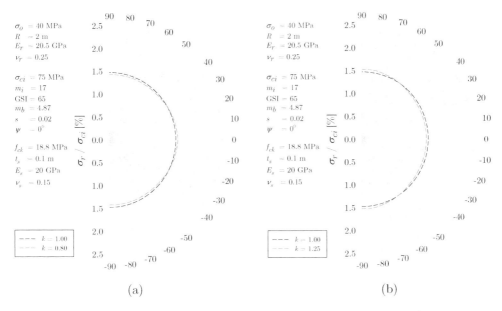

Fig. 5.8. Distributions of Radial Stresses along the Rock-Shotcrete Interface

In this chapter, unlike a plate element, the shotcrete was modelled using continuum elements, allowing a possibility to reveal the stresses in the shotcrete along its thickness. If one considers the three-dimensional tunnel advance and pre-relaxation ahead of the tunnel face when assessing the load transferred to the shotcrete subjected to non-uniform in-situ stresses, the use of three-dimensional models is recommended for acquiring more accurate results.

5.5. Plastic Zone

The plastic-elastic interface in the rock mass when the horizontal-to-vertical stress coefficient, k, equals to 1.00, 0.80 and 1.25 is presented in Fig. 5.9. While Fig. 5.9a shows the plastic zone, R_{pl}, before the shotcrete installation, Fig. 5.9b illustrates the plastic zone after the shotcrete installation.

In contrast to cases when the in-situ stresses in the rock mass are uniform ($k = 1$), the radius of the plastic-elastic interface for cases when the in-situ stresses are non-uniform ($k \neq 1$), is not constant. If $k < 1$, the shape of the plastic zone is oval with its major axis perpendicular to the direction of in-situ vertical stress, while if $k > 1$, an oval-shaped plastic zone is also developed but its major axis is parallel to the direction of in-situ vertical stress. As illustrated in Fig. 5.9a and b, the extent of the plastic zone can be reduced by placing shotcrete. For $k = 0.80$, the radius of plastic-elastic interface at the sidewalls was minimized from $1.27R$ to $1.23R$, whereas at the roof and invert it was decreased from $1.18R$ to $1.14R$. For $k = 1.25$, reverse values from those predicted for $k = 0.80$ were acquired. Beyond the plastic zone, the effects of excavation on stresses in the rock mass decrease considerably.

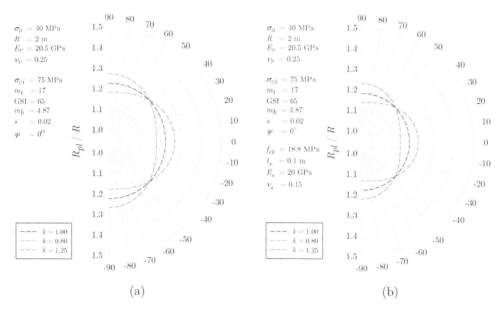

Fig. 5.9. Plastic-Elastic Interface: (a) Before, and (b) After the Shotcrete Installation

The proper assessment of plastic zone around a pressure tunnel is important for the design of support system. For stability purposes, the rock mass inside the plastic zone needs to be supported and the extent of the plastic-elastic interface in the rock mass determines the required length of rock bolts.

5.6. Prestress-Induced Hoop Stress in the Final Lining

After the completion of consolidation grouting, a final lining can be installed onto the shotcrete. Its thickness should satisfy the minimum lining thickness requirements to withstand the load imposed not only by the grouting pressure, but also, if any, by the external pressure such as groundwater above the tunnel. In practice, the thickness of a final lining, t_c, varies between 30 and 40 cm. Thinner concrete linings are preferable, since they will result in higher prestress-induced hoop stresses when compared to thicker concrete linings (Wannenmacher et al., 2012).

To prestress the final lining, cement-based grout is injected at high pressure into the circumferential gap between the shotcrete and the final lining. Thereby, the final lining is prestressed against the surrounding rock and a tight contact between the final lining, the shotcrete and the rock mass is achieved. The final lining, together with the rock mass, has to withstand the internal water pressure during tunnel operation. Here, the in-situ stresses are relevant regarding the bearing capacity of the tunnel.

For hydropower tunnels, the grouting pressure of up to 30 bar (3 MPa) is common to prestress a final lining. However, considering the strain losses due to creep, shrinkage and temperature change, it can be assumed that only 20 bar (2 MPa) of prestress, p_p, remains active at the shotcrete-final lining interface.

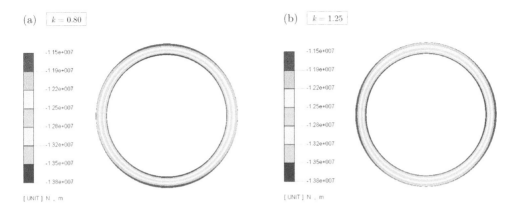

Fig. 5.10. Prestress-Induced Hoop Stresses in the Final Lining

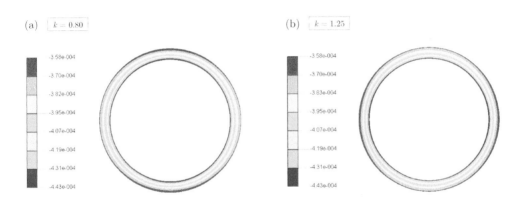

Fig. 5.11. Prestress-Induced Hoop Strains in the Final Lining

As an example, the concrete properties of C25/30 (Table 5.2) with reference to ÖNORM (2001) can be used as a data set for the final lining. To prestress the final lining, a uniform grouting pressure can be applied at the shotcrete-final lining interface. However, to avoid the final lining being influenced by the previous deformations during grouting, the radial deformations at the shotcrete-final lining interface has to be set to zero before activating the load representing the grouting pressure.

The numerical results of prestress-induced hoop stresses in a 30 cm final lining are shown in Fig. 5.10. The corresponding hoop strains are presented in Fig. 5.11. As a result of prestressing works, a slight degree of compressive hoop strains was induced throughout the final lining. If the in-situ vertical stress is greater than the in-situ horizontal stress or $k < 1$, the maximum prestress-induced hoop strain is located at the roof and invert of the final lining intrados. Conversely, if the in-situ horizontal stress is greater than the in-situ vertical stress or $k > 1$, the maximum prestress-induced hoop strain is found at the sidewalls of the final lining intrados.

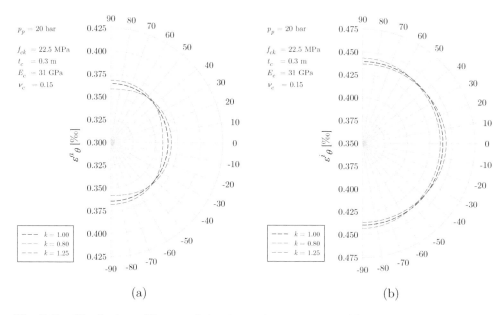

Fig. 5.12. Distributions of Prestress-Induced Hoop Strains along the (a) Extrados, and (b) Intrados of the Final Lining

The distributions of prestress-induced hoop strains along the extrados and intrados of the final lining are depicted in Fig. 5.12a and b, respectively. In particular, the hoop strains along the intrados of the final lining are of great interest since the assessment of the maximum internal water pressure is based on the lowest value of these strains.

For $k = 0.80$, the prestress-induced hoop strain at the roof and invert of the final lining intrados was found as 0.44‰, while at the sidewalls it was 0.43‰ (Fig. 5.12b). For $k = 1.25$, the prestress-induced hoop strain at the roof and invert of the final lining intrados was found as 0.43‰, while at the sidewalls it was 0.44‰ (Fig. 5.12b). Again, reverse values from those obtained for $k = 0.80$, were acquired for $k = 1.25$.

Since much of the tensile strength of concrete has already been used in the thermal cooling, the low tensile strength of concrete is neglected during the assessment of the maximum internal water pressure. For prestressed concrete-lined pressure tunnels, the assessment of the maximum internal water pressure can be done by offsetting the seepage-induced hoop strains at the final lining intrados against the prestress-induced hoop strains.

Moreover, the residual hoop stress in the final lining should be maintained in a compressive state of stress in order to avoid longitudinal cracks in the final lining and eventually hydraulic jacking of the surrounding rock mass. The criterion to assess the maximum internal water pressure for uncracked prestressed concrete-lined pressure tunnels can be written as follows:

$$\varepsilon^i_{\vartheta,\,p_p} + \varepsilon^i_{\vartheta,\,p_i} \leq 0 \tag{5.5}$$

in which $\varepsilon^i_{\vartheta,\ p_p}$ and $\varepsilon^i_{\vartheta,\ p_i}$ denote the prestress-induced hoop strain and the seepage-induced hoop strain at the final lining intrados, respectively.

According to Eq. (5.5), as high as 234 m of static water head, which corresponds to a 23 bar (2.3 MPa) of internal water pressure, p_i, was activated inside the pressure tunnel. The numerical results of radial stresses and hoop strains in the final lining for both $k = 0.80$ and $k = 1.25$, are presented in Figs. 5.13 and 5.14 respectively.

As shown in Fig. 5.14, the compressive hoop strains along the final lining intrados decreased to zero when a 234 m static water level was activated. Beyond this level, the final lining falls in a tensile state of stress and is vulnerable to cracking. Therefore, it is important to emphasize that once the maximum internal water pressure is obtained, a certain factor of safety should be applied before applying the predicted value into practice.

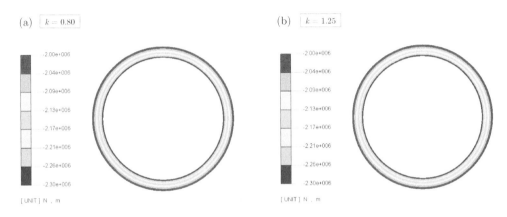

Fig. 5.13. Radial Stresses in the Final Lining during the Tunnel Operation

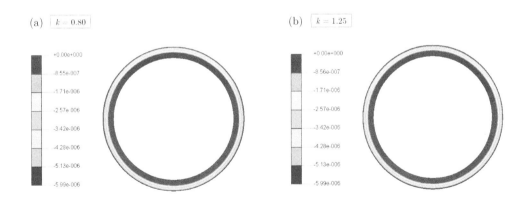

Fig. 5.14. Hoop Stresses in the Final Lining during the Tunnel Operation

As a result of the activation of a 234 m static water head, it can be seen that the residual hoop strains along the final lining extrados still remained in a compressive state of stress (Fig. 5.15). For $k = 0.80$, the residual hoop strain at the roof and invert of the final lining extrados was found as 5.74×10^{-6}, whereas at the sidewalls it was 5.97×10^{-6}. It can be seen that reverse values from those obtained for $k = 0.80$ were acquired for $k = 1.25$ (Fig. 5.15).

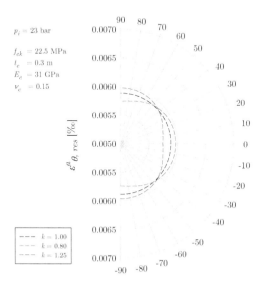

Fig. 5.15. Distribution of Residual Hoop Strains along the Final Lining Extrados

Once the hoop stress in the final lining during tunnel operation exceeds the tensile strength of concrete, utilizing the information given in Fig. 5.12, areas in the final lining where longitudinal cracks are likely to occur, can be identified. If the in-situ vertical stress in the rock mass is greater than the in-situ horizontal stress, longitudinal cracks can occur at the sidewalls of the final lining since the prestress-induced hoop strain at the sidewalls is smaller than those at any other locations along the inner perimeter of the final lining. On the contrary, if the in-situ horizontal stress is greater than the in-situ vertical stress, longitudinal cracks can occur at the roof and invert of the final lining.

5.7. Seepage Pressure around a Pressure Tunnel

In view of pervious concrete linings, seepage into the rock mass needs to be investigated. If the safety of pressure tunnel is not put at risks, seepage, q, up to 2 l/s/bar per km length of the tunnel is still tolerable (Marence, 2008). Otherwise, seepage in the order of 1 l/s/bar per km (Schleiss, 1988; 2013) should not be exceeded in order to avoid hydraulic jacking of the surrounding rock mass, washing out of joint fillings and associated hazards such as landslide, environmental impacts, flooding of the adjacent powerhouse and collapse of the tunnel.

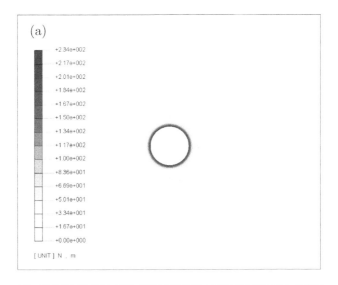

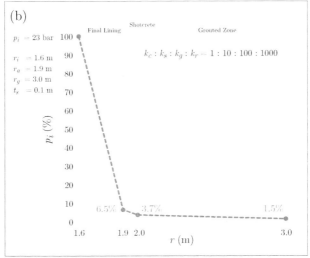

Fig. 5.16. (a) Pore Pressure Head and (b) Seepage-Induced Stresses around the Pressure Tunnel

The seepage around a prestressed concrete-lined pressure tunnel can be determined based on the distribution of seepage pressures through the final lining, the shotcrete, the grouted zone and the rock mass. In this chapter, the permeability coefficients for the rock mass, k_r, the grouted zone, k_g, the shotcrete, k_s, and the final lining, k_c, were taken as 10^{-6} m/s, 10^{-7} m/s, 10^{-8} m/s and 10^{-9} m/s, respectively. To obtain the pore pressure head distribution around the tunnel, the groundwater flow stress analysis in DIANA was employed.

The distribution of pore pressure head around the tunnel is depicted in Fig. 5.16a. As a result of the activation of a 234 m static water head, as much as 6.5% of the internal water pressure was transmitted to the final lining extrados (Fig. 5.16b).

The relative seepage pressures at the shotcrete-grouted zone interface, p_s, and at the grouted zone-rock mass interface, p_g, were found as 3.7% and 1.5%, respectively (Fig. 5.16b). Since the seepage pressure behind the final lining is still lower than the smallest in-situ stress in the rock mass, the tunnel stability against hydraulic jacking can be preserved.

5.8. Concluding Remarks

The horizontal-to-vertical stress coefficients, k, is one of the most important aspects in the design of hydropower tunnels. The primary focus in this chapter is to investigate the mechanical and hydraulic behaviour of prestressed concrete-lined pressure tunnels embedded in elasto-plastic isotropic rock mass whose in-situ stresses are non-uniform.

By means of a two-dimensional finite element model, the load sharing between the rock mass and the final lining can be revealed. This research shows that the in-situ stress ratio affects the distribution of stresses and deformations, the development of plastic zone, and thus the bearing capacity of pressure tunnels.

The maximum internal water pressure of prestressed concrete-lined pressure tunnels embedded in elasto-plastic rock mass whose in-situ stresses are non-uniform, can be assessed by offsetting the seepage-induced hoop strains at the final lining intrados against the prestress-induced hoop strains. As well as the internal water pressure, this approach is also useful to identify areas where longitudinal cracks may occur in the final lining. If the in-situ vertical stress in the rock mass is greater than the in-situ horizontal stress, longitudinal cracks can occur at the sidewalls. In contrast, if the in-situ horizontal stress is greater than the in-situ vertical stress, longitudinal cracks at the roof or invert of tunnels may be expected.

The mechanical and hydraulic behaviour of pressure tunnels presented herein was investigated based on the main assumption of isotropic rock mass, provided that there is no preferred orientation of joints within the rock mass. If the rock mass supporting the pressure tunnel exhibits significant anisotropy in strength and deformability, a new approach needs to be developed in order to account for the effects of stratification in rocks on the tunnel bearing capacity as well as to assess seepage into the rock mass.

6 | Pressure Tunnels in Transversely Isotropic Rock Formations[4,5]

This chapter deals with the behaviour of prestressed concrete-lined pressure tunnels embedded in transversely isotropic rocks subjected to either uniform or non-uniform in-situ stresses. A two-dimensional plane strain finite element model was employed to investigate the effect of the orientation of stratifications in the rock mass, also known as the dip angle, a, and the horizontal-to-vertical stress coefficient, k, on the bearing capacity of such tunnels.

As long as the in-situ stresses in the rock mass are uniform, the load sharing between the rock mass and the final lining is influenced by the dip angle. Otherwise, it will be influenced not only by the dip angle but also by the in-situ stress ratio. If the in-situ stresses in the rock mass are uniform, the distribution of stresses and deformations as a result of the tunnel construction process demonstrates a symmetrical pattern to the orientation of stratification planes. If the in-situ stresses are non-uniform, the distribution of stresses and deformations exhibits an unsymmetrical pattern for cases of tunnels embedded in transversely isotropic rocks with inclined stratification planes. This research suggests that the distribution of stresses and deformations obtained for a specific value a with coefficient k is identical to that for $a + 90°$ with coefficient $1/k$ by rotating the tunnel axis by $90°$.

The bearing capacity of prestressed concrete-lined pressure tunnels was determined based on the superposition principle of hoop strains at the final lining intrados. As the maximum internal water pressure, potential locations where longitudinal cracks may occur in a concrete lining prestressed in transversely isotropic rocks can also be identified.

---

[4] Based on Simanjuntak, T.D.Y.F., Marence, M., Mynett, A.E., Schleiss, A.J. (2014). *Effects of Rock Mass Anisotropy on Deformations and Stresses around Tunnels during Excavation*. The 82nd Annual Meeting of ICOLD, International Symposium on Dams in a Global Environmental Challenges. 01-06 June 2014. Bali, Indonesia, pp. II-129 – II-136.

[5] Based on Simanjuntak, T.D.Y.F., Marence, M., Schleiss, A.J., Mynett, A.E. (2015). *The Interplay of In-Situ Stress Ratio and Transverse Isotropy in the Rock Mass on Prestressed Concrete-Lined Pressure Tunnels*. Tunnelling and Underground Space Technology (Under Review).

6.1. Introduction

Most designs of concrete-lined pressure tunnels consider the rock mass supporting the tunnel as an isotropic material (Seeber, 1985a; 1985b; Schleiss, 1986b; Simanjuntak et al., 2012a; Wannenmacher et al., 2012), which is usually acceptable given that the rock mass exhibits non-significant anisotropy in strength and deformability. This consideration has contributed not only to the knowledge of the mechanical-hydraulic interaction between the final lining and the rock mass, but also to the investigation of the behaviour of prestressed concrete-lined pressure tunnels embedded in elasto-plastic isotropic rock mass subjected to uniform (Simanjuntak et al., 2012b) as well as to non-uniform (Simanjuntak et al., 2014c) in-situ stresses.

Pressure tunnels, nevertheless, may be constructed in an inherently anisotropic rock mass, such as metamorphic rocks. These types of rocks, which are composed of lamination of intact rocks, can take the form of cross anisotropy or transverse isotropy commonly configured by one direction of lamination plane perpendicular to the direction of deposition (Gao et al., 2010). In such cases, the rock supporting the tunnel may exhibit significant strength and deformability in the direction parallel and perpendicular to the stratification planes; rendering the behaviour of pressure tunnels embedded in such rocks can deviate from that investigated under the assumption of isotropic rocks.

Aside from the in-situ stress ratio, the orientation of stratification planes referred to as the dip angle is yet another aspect that has to be considered in the design of pressure tunnels embedded in transversely isotropic rocks. It may influence the bearing capacity of the tunnel since the distribution of stresses and displacements around the tunnel depends on the direction-dependent properties of the rock mass.

By employing a two-dimensional finite element model, the mechanical and hydraulic behaviour of prestressed concrete-lined pressure tunnels in transversely isotropic rocks can be investigated. The rock mass being considered has either uniform or non-uniform in-situ stresses. To allow for the use of a two-dimensional model, the pressure tunnel examined is assumed to be driven in the direction parallel to the strike of the planes of transverse isotropy. The interplay between the transverse isotropy and the in-situ stress ratio on the tunnel lining performance is explored based on the concept that there is no slip allowed to occur along the stratification planes. In accordance with the previous chapters, this chapter concentrates on a deep, straight ahead circular tunnel situated above the groundwater level.

In this chapter, the analysis begins with the response of the rock mass to excavation and continues to the prediction of stresses and deformations as a result of simultaneous excavation and support installation. A final lining is installed onto the shotcrete and prestressed by using the passive prestressing technique. The maximum internal water pressure is assessed by offsetting the seepage-induced hoop strains at the final lining intrados against the prestress-induced hoop strains. Finally, locations where longitudinal cracks can occur in the final lining are identified.

6.2. Tunnel Excavation in Transversely Isotropic Rocks

As long as discontinuities in the rock mass are more or less parallel and regularly spaced, the rock mass can at first be approximated as elastic transversely isotropic material. The influence of stratification on the behaviour of the rock mass can be considered by incorporating different deformability properties at directions parallel and perpendicular to the surface of dominant discontinuities (Fortsakis et al., 2012; Kolymbas et al., 2012).

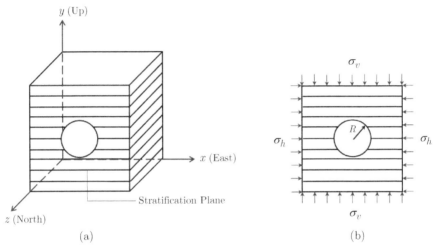

(a)   (b)

Fig. 6.1. Circular Tunnel Driven in Transversely Isotropic Rocks Having Horizontal Stratification Planes

Fig. 6.1 illustrates a representation of problems of a circular underground opening in transversely isotropic rocks with horizontal stratification planes, i.e. $a = 0°$. While the z-axis is the tunnel axis, the x- and z-axis are the plane of transverse isotropy (Fig. 6.1a). Generally, the in-situ stresses in a rock mass are non-uniform (Fig. 6.1b). The in-situ horizontal stress, σ_h, can be expressed as the product of the in-situ vertical stress, σ_v, and a coefficient of earth pressure, k.

$$\sigma_h = k \, \sigma_v \tag{6.1}$$

The mean in-situ stress, σ_o, can be defined as:

$$\sigma_o = \frac{\sigma_v + \sigma_h}{2} = \frac{(k+1)\,\sigma_v}{2} \tag{6.2}$$

in which the in-situ stresses are non-uniform, if $k \neq 1$.

As long as the plane of transverse isotropy strikes parallel to the tunnel axis, plane strain conditions apply along the tunnel axis and the components ε_z, ε_{yz}, and ε_{xz} vanish everywhere. The constitutive model in plane strain conditions are given as:

$$\begin{pmatrix} \varepsilon_x \\ \varepsilon_y \\ \varepsilon_{xy} \end{pmatrix} = \begin{pmatrix} S_{11} & S_{12} & 0 \\ S_{21} & S_{22} & 0 \\ 0 & 0 & S_{33} \end{pmatrix} \begin{pmatrix} \sigma_x \\ \sigma_y \\ \tau_{xy} \end{pmatrix} \tag{6.3}$$

where σ_x and σ_y are the total stress along the x- and y-axis respectively, τ_{xy} is the shear stress, S_{11}, S_{21}, S_{12}, S_{22}, and S_{33} are the compliance coefficients and related to the material parameters as follows:

$$
\begin{aligned}
S_{11} &= \frac{1 - \nu_h^2}{E_h} \\[2mm]
S_{22} &= \frac{1 - \nu_{hv}\,\nu_{vh}}{E_v} \\[2mm]
S_{12} &= S_{21} = -\frac{\nu_{vh}\,(1 + \nu_h)}{E_v} \\[2mm]
S_{33} &= \frac{1}{G_{vh}}
\end{aligned}
\tag{6.4}
$$

in which E_h and E_v are the Young's modulus in the plane of isotropy and in the direction normal to the plane of isotropy respectively, ν_h is the Poisson's ratio in the plane of isotropy, ν_{hv} is the Poisson's ratio for the effect of stress in the plane of isotropy on the strain in the direction normal to the plane of isotropy, ν_{vh} is the Poisson's ratio for the effect of stress normal to the plane of isotropy on the strain in the plane of isotropy, and G_{vh} is the shear modulus normal to the plane of isotropy.

A full mathematical treatise to determine excavation-induced hoop stresses and radial deformations along the perimeter of a circular tunnel embedded in elastic transversely isotropic rocks with horizontal stratification planes can be found in Hefny and Lo (1999). For completeness, the closed-form solutions are included herein. In relation to Eqs. (6.1) and (6.2), they can be written as:

Hoop stresses:

$$
\sigma_\vartheta = \frac{2 + 2(\gamma_1 + \gamma_2)^2 - 2\gamma_1^2\gamma_2^2 - 4(\gamma_1 + \gamma_2)\cos 2\vartheta}{(1 + \gamma_1^2 - 2\gamma_1\cos 2\vartheta)(1 + \gamma_2^2 - 2\gamma_2\cos 2\vartheta)}\left(\frac{(k+1)\sigma_v}{2}\right) + \tag{6.5}
$$
$$
+ \frac{4(\gamma_1 + \gamma_2) - 4(1 + \gamma_1\gamma_2)\cos 2\vartheta}{(1 + \gamma_1^2 - 2\gamma_1\cos 2\vartheta)(1 + \gamma_2^2 - 2\gamma_2\cos 2\vartheta)}\left(\frac{(k-1)\sigma_v}{2}\right)
$$

with

$$\gamma_1 = \frac{\mu_1 - 1}{\mu_1 + 1}; \quad |\gamma_1| < 1$$

$$\gamma_2 = \frac{\mu_2 - 1}{\mu_2 + 1}; \quad |\gamma_2| < 1$$

$$\mu_1^2 \mu_2^2 = \frac{S_{11}}{S_{22}}$$

$$\mu_1^2 + \mu_2^2 = \frac{2 S_{12} + S_{33}}{S_{22}}$$

(6.6)

Radial deformations:

$$u_r = \frac{R}{2 (\gamma_1 - \gamma_2)} \cdot$$

(6.7)

$$\cdot \left\{ \begin{array}{l} \left[\left(\frac{(k+1)\, \sigma_v}{2} \right) (\gamma_2\, \varrho_1 - \gamma_1\, \varrho_2) + \left(\frac{(k-1)\, \sigma_v}{2} \right) (\varrho_1 - \varrho_2) + \right. \\ \left. + \left[\left(\frac{(k+1)\, \sigma_v}{2} \right) (\gamma_2\, \delta_1 - \gamma_1\, \delta_2) + \left(\frac{(k-1)\, \sigma_v}{2} \right) (\delta_1 - \delta_2) \right] \cos 2\vartheta \right. \end{array} \right\}$$

with

$$\delta_1 = (1 + \gamma_1)\, \beta_2 - (1 - \gamma_1)\, \beta_1$$
$$\delta_2 = (1 + \gamma_2)\, \beta_1 - (1 - \gamma_2)\, \beta_2$$
$$\varrho_1 = (1 + \gamma_1)\, \beta_2 + (1 - \gamma_1)\, \beta_1$$
$$\varrho_2 = (1 + \gamma_2)\, \beta_1 + (1 - \gamma_2)\, \beta_2$$

(6.8)

and

$$\beta_1 = S_{12} - S_{22}\, \mu_1^2$$
$$\beta_2 = S_{12} - S_{22}\, \mu_2^2$$

(6.9)

As discussed by Hefny and Lo (1999) and Manh et al. (2014), two cases may arise which are either all parameters δ_1, δ_2, ϱ_1, and ϱ_2 are real if γ_1 and γ_2 are real, or parameters δ_1 and δ_2 as well as ϱ_1 and ϱ_2 are complex conjugates if γ_1 and γ_2 are complex conjugates.

Fig. 6.2 represents a circular underground opening in transversely isotropic rocks with non-horizontal stratification planes, i.e. $a \neq 0°$. Consequently, all deformation components including those appearing in Eq. (6.7) depend also on the orientation of stratification planes, a. In general, the excavation-induced radial deformations, u_r, can be written in dimensionless form as:

$$\frac{u_r}{R} = f \left(\frac{E_h}{E_v}, \frac{G_{vh}}{E_h}, \frac{\nu_h}{\nu_{vh}}, k, a \right)$$

(6.10)

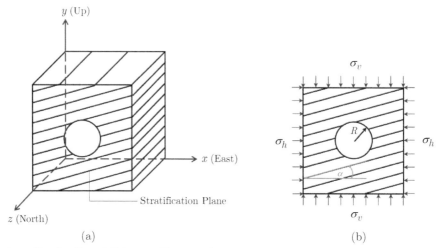

Fig. 6.2. Circular Tunnel Driven in Transversely Isotropic Rocks Having Non-Horizontal Stratification Planes

By means of an elasto-plastic Jointed Rock model implemented in the finite element code DIANA, an attempt was made herein to explore the effect of the orientation of stratification planes and the in-situ stress ratio on the elastic response of transversely isotropic rocks to the tunnel excavation. Providing that there is no slip between the planes of transverse isotropy, the rock mass as a whole can be idealized as a continuous material. In the model, the elastic response of transversely isotropic rocks can be ensured by providing an adequate cohesion along the sliding planes (Wittke, 1990; Tonon and Amadei, 2003; Tonon, 2004; Simanjuntak et al., 2014a).

Table 6.1. Rock Mass Properties (Hefny and Lo, 1999)

E_h (GPa)	E_v (GPa)	G_{vh} (GPa)	ν_{vh}	ν_h
15.8	10.5	3.95	0.30	0.30

As an example, the circular excavation with a radius of 2 m is executed to the rock mass having transversely isotropic formations. The rock mass properties are listed in Table 6.1 with reference to Hefny and Lo (1999). The mean in-situ stress in the rock mass, σ_o, is 40 MPa. The orientation of stratification planes being examined is 0°, 45°, 90°, and 135°, with 0° indicating horizontal stratification planes. In this research, the orientation of stratification planes is measured counterclockwise from the x-axis.

Regarding the in-situ stresses in the rock mass, two cases are distinguished based on whether the in-situ stresses are uniform ($k = 1$), or not ($k \neq 1$). For cases where the in-situ stresses are non-uniform, two distinctive sub-cases are studied, namely the case in which the in-situ vertical stress is greater than the horizontal ($k < 1$), and the other case in which the in-situ horizontal stress is greater than the vertical ($k > 1$). Since the tunnel stands in deep rock, the gravitational force is negligible (Detournay and Fairhurst, 1987).

Fig. 6.3 presents the numerical results of excavation-induced radial deformations in the space surrounding the tunnel subjected to uniform in-situ stresses ($k = 1$). It can be seen that the distribution of radial deformations shows a symmetrical pattern to the orientation of stratification planes when the in-situ stresses are uniform.

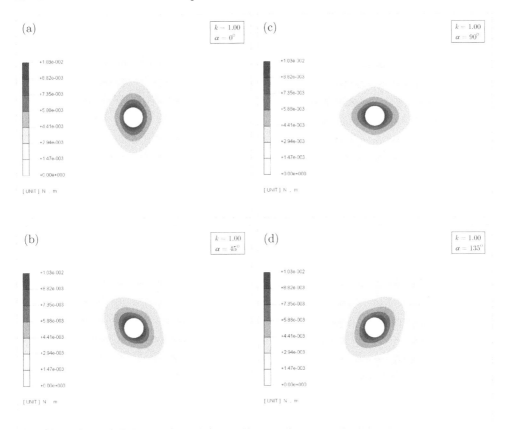

Fig. 6.3. Excavation-Induced Radial Deformations for Cases when the In-Situ Stresses in the Rock Mass are Uniform

The numerical results of excavation-induced radial deformations along the tunnel perimeter for $k = 1$ are depicted in Fig. 6.4a. If the stratification planes are horizontal, the corresponding radial convergence at the tunnel roof and invert was calculated as $u_r/2R = 0.26\%$ whereas at the sidewalls it was 0.20%. If the stratification planes are inclined at $45°$ above the horizontal, i.e. at $a = 45°$, the maximum deformation corresponding to the radial convergence equals to 0.26% was observed at the tunnel arcs, specifically at $135°$ and $315°$ counted counterclockwise from the x-axis, whereas the minimum deformation corresponding to the radial convergence equals to 0.20% was situated at $45°$ and $225°$. If the stratification planes are vertical, the maximum deformation corresponding to the radial convergence equals to 0.26% was observed at the tunnel sidewalls, whereas the minimum deformation corresponding to the radial convergence equals to 0.20% was found at the tunnel roof and invert. If the stratification planes are inclined at $135°$, the maximum deformation corresponding to the

radial convergence equals to 0.26% was found at 45° and 225° counted counterclockwise from the x-axis, while the minimum deformation corresponding to the radial
convergence equals to 0.20% was situated at 135° and 315°. These results are in good
agreement with those calculated using the closed-form solutions (Fig. 6.4a).

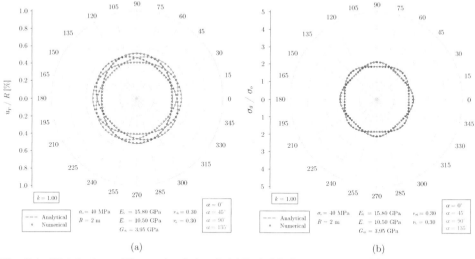

(a)   (b)

Fig. 6.4. Distributions of Excavation-Induced- (a) Radial Deformations, and (b) Hoop Stresses along
the Tunnel Perimeter for Cases when the In-Situ Stresses in the Rock Mass are Uniform

The numerical results of excavation-induced hoop stresses around the underground
opening for $k = 1$ are illustrated in Fig. 6.5, with a negative sign indicating a compressive state of stress. For various a, the excavation-induced hoop stresses along the
tunnel perimeter are depicted in Fig. 6.4b.

As shown in Fig. 6.4b, when the stratification planes are horizontal, the maximum
excavation-induced hoop stress, σ_θ, equals to $2.14\sigma_o$ was observed at the tunnel roof
and invert, whereas the minimum hoop stress equals to $1.86\sigma_o$ was found at the tunnel arcs, specifically at 45° and 225° measured counterclockwise from the x-axis. If
the stratification planes are vertical, the maximum hoop stress equals to $2.14\sigma_o$ was
found at the tunnel sidewalls, whereas the minimum hoop stress equals to $1.86\sigma_o$ was
noticed at 45° and 225°. If $a = 45°$, the maximum excavation-induced hoop stress
equals to $2.14\sigma_o$ was found at 135° and 315°, while the minimum hoop stress equals to
$1.86\sigma_o$ was located at the sidewalls. Moreover, if $a = 135°$, the maximum excavation-
induced hoop stress equals to $2.14\sigma_o$ was found at 45° and 225°, while the minimum
hoop stress equals to $1.86\sigma_o$ was located at the sidewalls. As shown in Fig. 6.4b, the
good agreement between the numerical results and those calculated using the closed-
form solutions is evident.

Based on the results depicted in Fig. 6.4a and b, it can be seen that the distribution
of excavation-induced radial deformations as well as hoop stresses obtained for a specific value a is identical to that for $a + 90°$, by rotating the tunnel axis by 90°.

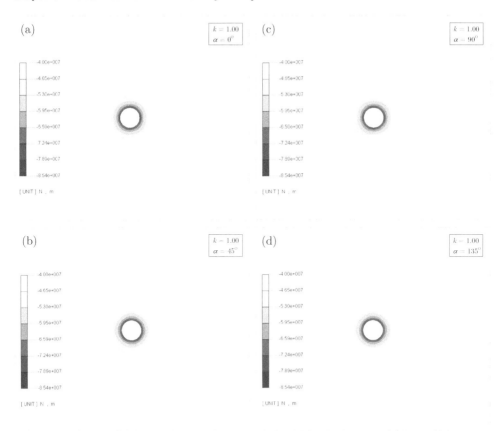

Fig. 6.5. Excavation-Induced Hoop Stresses for Cases when the In-Situ Stresses in the Rock Mass are Uniform

Fig. 6.6 shows the numerical results of excavation-induced radial deformations in the space surrounding transversely isotropic rocks subjected to non-uniform in-situ stresses ($k \neq 1$). While Fig. 6.6a, b and c presents the predicted excavation-induced radial deformations for $k = 0.80$, Fig. 6.6d, e and f depicts the results for $k = 1.25$.

As reported by Hefny and Lo (1999), the distribution of excavation-induced radial deformations in the space surrounding the tunnel exhibits a symmetrical pattern for cases where the stratification planes are horizontal. Also, the distribution of radial deformations is symmetrical for cases when the stratification planes are vertical, as shown in Fig. 6.6b and e.

If the stratification planes are inclined (Fig. 6.6a, c, d and f), the distribution of radial deformations demonstrates an unsymmetrical pattern. Interestingly, comparing Fig. 6.6a with f or Fig. 6.6c with d, it is seen that the distribution of excavation-induced radial deformations for a specific value a with coefficient k is identical to that for $a + 90°$ with coefficient $1/k$, by rotating the tunnel axis by 90°.

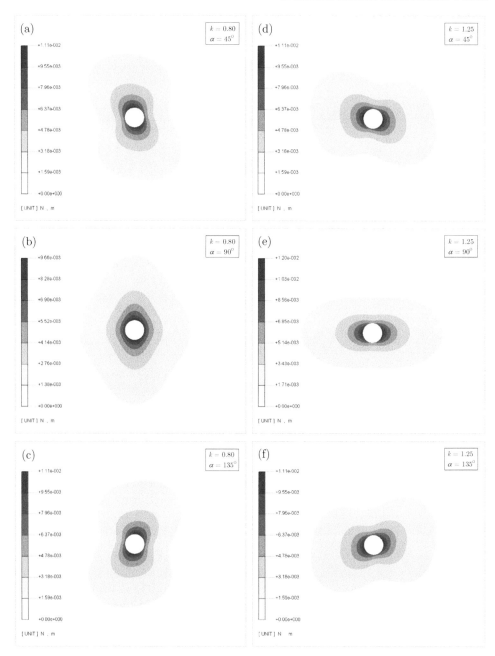

Fig. 6.6. Excavation-Induced Radial Deformations for Cases when the In-Situ Stresses in the Rock Mass are Non-Uniform

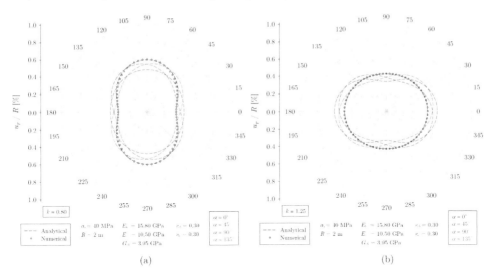

Fig. 6.7. Distributions of Excavation-Induced Radial Deformations along the Tunnel Perimeter for Cases when the In-Situ Stresses in the Rock Mass are Non-Uniform

The predicted excavation-induced radial deformations along the tunnel perimeter are illustrated in Fig. 6.7. In respect of accuracy, the numerical results for cases of tunnels where $a = 0°$ are compared with those calculated by using the closed-form solutions. It can be seen that the numerical results obtained for $a = 0°$ for both $k = 0.80$ (Fig. 6.7a) and $k = 1.25$ (Fig. 6.7b) conform to the analytical results; implying that the numerical model can reproduce the response of transversely isotropic rocks in a realistic manner and can be applied to obtain results for $a = 45°$, $90°$ and $135°$.

In the following, the numerical results of excavation-induced radial deformations for cases where the in-situ vertical stress is greater than the in-situ horizontal stress, i.e. $k < 1$, are analysed. For a ratio $k = 0.80$ (Fig. 6.7a) and if the stratification planes are horizontal, the corresponding radial convergence at the tunnel roof and invert was calculated as 0.30%, whereas at the sidewalls it was 0.17%. If the stratification planes are inclined at 45°, or at $a = 45°$, the maximum deformation corresponding to the radial convergence equals to 0.28% was found at the tunnel arcs, specifically at 105° and 285° counted counterclockwise from the x-axis, whereas the minimum deformation corresponding to the radial convergence equals to 0.18% was located at 15° and 195°. If the stratification planes are vertical, the maximum deformation corresponding to the radial convergence equals to 0.24% was observed at the tunnel roof and invert, while the minimum deformation corresponding to the radial convergence equals to 0.22% was found at the sidewalls. If the stratification planes are inclined at 135°, the maximum deformation corresponding to the radial convergence equals to 0.28% was found at the tunnel arcs, specifically at 75° and 255° counted counterclockwise from the x-axis, whereas the minimum deformation corresponding to the radial convergence equals to 0.18% was located at 165° and 345°.

The results in Fig. 6.7a suggest that the shape of the underground opening is oval with its major axis parallel to the direction of stratification planes in the rock mass. As long as the in-situ vertical stress is greater than the in-situ horizontal stress, the ovalization of the tunnel decreases as the dip angle increases.

Similarly, the response of transversely isotropic rocks to a circular tunnel excavation was also investigated for cases where the in-situ horizontal stress is greater than the in-situ vertical stress, i.e. $k > 1$. For a ratio $k = 1.25$ (Fig. 6.7b) and if the stratification planes are horizontal, the maximum deformation corresponding to the radial convergence equals to 0.24% was found at the sidewalls, while the minimum deformation corresponding to the radial convergence equals to 0.22% was located at the roof and invert of the tunnel; implying that the shape of the underground opening is also oval with its major axis perpendicular to the direction of stratification planes. If $a = 45°$, the maximum deformation corresponding to the radial convergence equals to 0.28% was found at 165° and 345°, while the minimum deformation corresponding to the radial convergence equals to 0.18% was situated at 75° and 255°. If the stratification planes are vertical, the maximum deformation corresponding to the radial convergence equals to 0.30% was found at the roof and invert, while the minimum deformation corresponding to the radial convergence equals to 0.17% was obtained at the sidewalls. If $a = 135°$, the maximum deformation corresponding to the radial convergence equals to 0.28% was found at 15° and 195°, while the minimum deformation corresponding to the radial convergence equals to 0.18% was situated at 105° and 285°. The results in Fig. 6.7b suggest that the ovalization of the tunnel increases as the dip angle increases, if the in-situ horizontal stress in the rock mass is greater than the in-situ vertical stress.

The numerical results of excavation-induced hoop stresses surrounding the tunnel subjected to non-uniform in-situ stresses ($k \neq 1$) are depicted in Fig. 6.8. As comparison, the analytical results of excavation-induced hoop stresses along the tunnel perimeter for cases where the stratification planes are horizontal are presented in Fig. 6.9. It can be seen that the good agreement between the numerical and analytical results is evident.

Identical to that observed by Hefny and Lo (1999), the distribution of excavation-induced hoop stresses surrounding the tunnel demonstrates a symmetrical pattern if the stratification planes are horizontal. The distribution of hoop stresses illustrates also a symmetrical pattern if the stratification planes are vertical (Fig. 6.9a and b).

For cases where the in-situ vertical stress is greater than the in-situ horizontal stress, i.e. $k < 1$, the maximum hoop stress is located at the tunnel sidewalls, whereas the minimum hoop stress is observed at the tunnel roof and invert. Conversely, for cases where the in-situ horizontal stress is greater than the vertical, i.e. $k > 1$, the maximum hoop stress is located at the tunnel roof and invert, while the minimum hoop stress is noticed at the tunnel sidewalls.

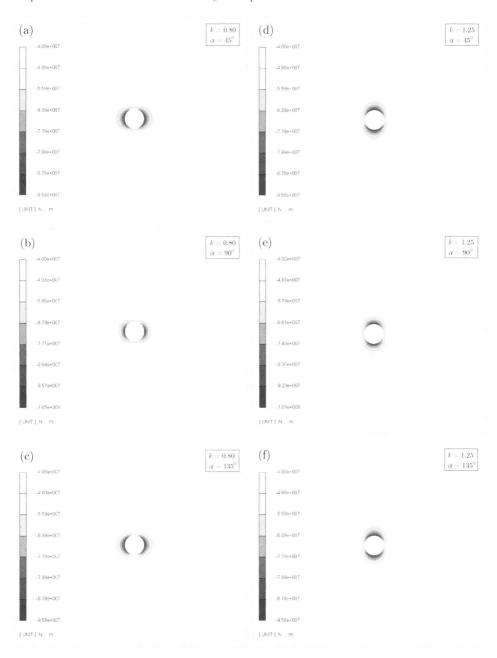

Fig. 6.8. Excavation-Induced Hoop Stresses for Cases when the In-Situ Stresses in the Rock Mass are Non-Uniform

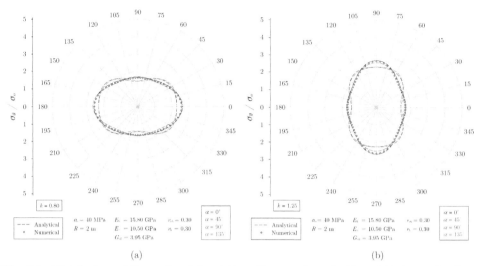

(a)   (b)

Fig. 6.9. Distributions of Excavation-Induced Hoop Stresses along the Tunnel Perimeter for Cases when the In-Situ Stresses in the Rock Mass are Non-Uniform

For $k = 0.80$ (Fig. 6.9a) and if the stratification planes are horizontal, the excavation-induced hoop stress at the tunnel sidewalls was found as $2.55\sigma_o$, while at the tunnel roof and invert it was $1.65\sigma_o$. If the stratification planes are vertical, the excavation-induced hoop stress at the sidewalls was obtained as $2.66\sigma_o$, whereas at the roof and invert it was $1.70\sigma_o$. For $k = 1.25$ (Fig. 6.9b) and if the stratification planes are horizontal, the excavation-induced hoop stress at the roof and invert was found as $2.66\sigma_o$, while at the sidewalls it was $1.70\sigma_o$. If the stratification planes are vertical, the excavation-induced hoop stress at the roof and invert was found as $2.55\sigma_o$, while at the sidewalls it was $1.65\sigma_o$.

Also shown in Fig. 6.9a and b, the distribution of hoop stresses illustrates an unsymmetrical pattern for tunnels driven in transversely isotropic rocks whose stratification planes are inclined. When $k = 0.8$ (Fig. 6.9a) and if $a = 45°$, the maximum excavation-induced hoop stress equals to $2.34\sigma_o$ was situated at the tunnel arcs, specifically at $155°$ and $335°$ measured counterclockwise from the x-axis, whereas the minimum hoop stress equals to $1.43\sigma_o$ was observed at $95°$ and $275°$. Under the same loading, the maximum hoop stress equals to $2.34\sigma_o$ was found at $25°$ and $205°$, while the minimum hoop stress equals to $1.43\sigma_o$ was situated at $85°$ and $265°$ if $a = 135°$. When $k = 1.25$ (Fig. 6.9b) and if $a = 45°$, the maximum excavation-induced hoop stress equals to $2.34\sigma_o$ was observed at $115°$ and $295°$, whereas the minimum excavation-induced hoop stress equals to $1.43\sigma_o$ was found at $175°$ and $355°$. If $a = 135°$, the maximum hoop stress equals to $2.34\sigma_o$ was observed at $65°$ and $245°$, while the minimum hoop stress equals to $1.43\sigma_o$ was found at $5°$ and $185°$. These results imply that the distribution of excavation-induced hoop stresses for a specific value a with coefficient k is identical to that for $a + 90°$ with coefficient $1/k$, by rotating the tunnel axis $90°$.

Table 6.2. Predicted Excavation-Induced Radial Deformations and Hoop Stresses along the Tunnel Perimeter for cases where the In Situ Stresses are Uniform and Non-Uniform

k	a	$u_r/2R$ (%)				σ_θ/σ_o			
		Max	ϑ	Min	ϑ	Max	ϑ	Min	ϑ
0.80	0°	0.30	90°, 270°	0.17	0°, 180°	2.55	0°, 180°	1.65	90°, 270°
	45°	0.28	105°, 285°	0.18	15°, 195°	2.34	155°, 335°	1.43	95°, 275°
	90°	0.24	90°, 180°	0.22	0°, 180°	2.66	0°, 180°	1.70	90°, 270°
	135°	0.28	75°, 255°	0.18	165°, 345°	2.34	25°, 205°	1.43	85°, 265°
1.00	0°	0.26	90°, 270°	0.20	0°, 180°	2.14	90°, 270°	1.86	45°, 225°
	45°	0.26	135°, 315°	0.20	45°, 225°	2.14	135°, 315°	1.86	0°, 180°
	90°	0.26	0°, 180°	0.20	90°, 270°	2.14	0°, 180°	1.86	45°, 225°
	135°	0.26	45°, 225°	0.20	135°, 315°	2.14	45°, 225°	1.86	0°, 180°
1.25	0°	0.24	0°, 180°	0.22	90°, 270°	2.66	90°, 270°	1.70	0°, 180°
	45°	0.28	165°, 345°	0.18	75°, 255°	2.34	115°, 295°	1.43	175°, 355°
	90°	0.30	90°, 270°	0.17	0°, 180°	2.55	90°, 270°	1.65	0°, 180°
	135°	0.28	15°, 195°	0.18	105°, 285°	2.34	65°, 245°	1.43	5°, 185°

The numerical results of excavation-induced radial deformations and hoop stresses along the tunnel perimeter for cases when either the in-situ stresses in the rock mass are uniform ($k = 1$) or non-uniform ($k \neq 1$), are summarized in Table 6.2. If compared to the results when $k = 1$, it is seen that the horizontal-to-vertical stress coefficient, k, and dip angle, a, significantly influence the ovalization of the tunnel and the hoop stresses in the rock mass when $k \neq 1$. The hoop stress in particular, the maximum hoop stress increases and the minimum hoop stress decreases as the in-situ stresses are non-uniform.

According to the results presented in Table 6.2, it can be concluded that the rock mass is stable since the hoop stress in the rock mass still remained in a compressive state after the tunnel excavation. If the hoop stresses reach the tensile strength of the rock, the separation of the planes of transverse isotropy within the rock mass can occur and endanger the stability of the rock mass around the underground opening. As a result, roof fall-related problems may take place at the locations where the hoop stresses are low or in a tensile state of stress. Comparing Fig. 6.9a with b, it can be seen that the rock mass with inclined stratification planes is more vulnerable to such problems.

6.3. Radial Stresses and Deformations Transmitted to a Support System

Since it first appeared in Hoek and Marinos (2000), the critical strain concept that is a ratio between the tunnel closure and tunnel diameter, has been used as an indicator to assess potential tunnel instabilities, as well as guidelines to design a support system. For hydropower tunnels, shotcrete with a thickness of 5 to 10 cm is commonly applied to support the excavation as long as the critical strain is less than 1%. Otherwise, appropriate tunnel stabilization measures (Hoek and Marinos, 2000; Barla et al., 2011) have to be implemented to prevent large tunnel deformations due to the squeezing of rocks.

In this chapter, a 10 cm shotcrete having mechanical properties according to the concrete type C20/25 listed in Table 6.3 was used as an example. Regarding radial stresses and deformations at the rock-shotcrete interface, the same setting suggested by Bobet (2011), i.e. elastic response of the rock and shotcrete, tight contact between rock and shotcrete, two-dimensional plane strain conditions along the tunnel axis, and simultaneous tunnel excavation and shotcrete installation, was adopted.

Table 6.3. Concrete Properties (ÖNORM, 2001)

Type	E (GPa)	ν	f_{ctm} (MPa)	f_{ctk} (MPa)	f_{cwk} (MPa)	f_{ck} (MPa)
C20/25	20	0.15	2.2	1.5	25	18.8
C25/30	31	0.15	2.6	1.8	30	22.5

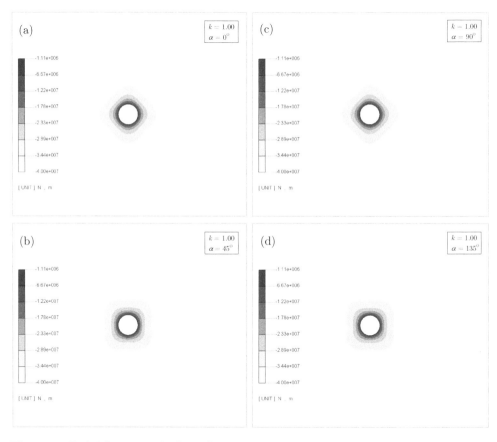

Fig. 6.10. Radial Stresses in the Space Surrounding the Tunnel after the Shotcrete Installation for Cases where the In-Situ Stresses in the Rock Mass are Uniform

The above assumptions can result in an overestimation of stresses transmitted to the shotcrete since the three-dimensional effect of excavation is not considered. However, the distribution of radial stresses and deformations along the thickness of the shotcrete can be revealed since the shotcrete is modelled by using two-dimensional plane strain continuum elements (Bonini et al., 2013; Simanjuntak et al., 2014c).

Fig. 6.10 presents the numerical results of radial stresses in the space surrounding the tunnel after the shotcrete installation for cases where the in-situ stresses in the rock mass are uniform ($k = 1$). It can be seen that the distribution of radial stresses shows a symmetrical pattern to the orientation of stratification planes if the in-situ stresses are uniform. The numerical results of radial deformations and radial stresses along the rock-shotcrete interface are depicted in Fig. 6.11a and b, respectively.

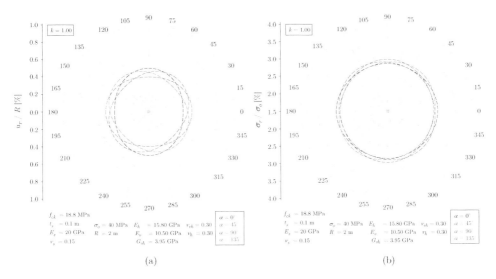

(a)   (b)

Fig. 6.11.   Distributions of (a) Radial Deformations, and (b) Hoop Stresses along the Rock-Shotcrete Interface when the In-Situ Stresses in the Rock Mass are Uniform

In the following, the numerical results of radial deformations (Fig. 6.11a) along the rock-shotcrete interface for cases where the in-situ stresses are uniform ($k = 1$) are analysed. If $a = 0°$, the radial convergence at the tunnel roof and invert was reduced to 0.25%, while at the sidewalls it was 0.20%. If $a = 45°$, the radial convergence at the tunnel arcs situated at 45° and 225° counted counterclockwise from the x-axis, was decreased to 0.25%, while the radial convergence at 135° and 315° was found as 0.20%. If $a = 90°$, the radial convergence at the tunnel roof and invert was found as 0.20%, whereas at the sidewalls it was reduced to 0.25%. Furthermore, if $a = 135°$, the radial convergence at 135° and 315° was decreased to 0.25%, while at 45° and 225° it was found as 0.20%.

Concerning the radial stresses (Fig. 6.11b) and if $a = 0°$, the maximum scaled radial stress, σ_r/σ_o, equals to 2.98% was observed at the sidewalls, while the minimum scaled radial stress equals to 2.86% was found at the roof and invert. Inverse results were found if $a = 90°$. If $a = 45°$, the maximum scaled radial stress equals to 2.98% was obtained at 135° and 315° counted counterclockwise from the x-axis, the minimum scaled radial stress equals to 2.86% was located at 45° and 225°. If $a = 135°$, the maximum scaled radial stress equals to 2.98% was obtained at 45° and 225°, while the minimum scaled radial stress equals to 2.86% was noticed at 135° and 315°.

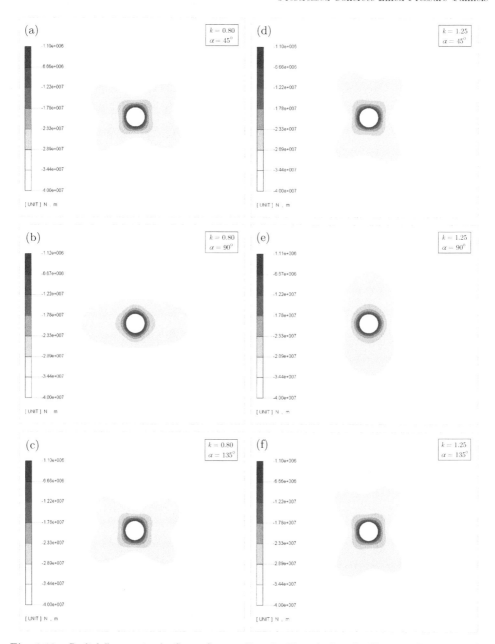

Fig. 6.12. Radial Stresses in the Space Surrounding the Tunnel after the Shotcrete Installation for Cases where the In-Situ Stresses in the Rock Mass are Non-Uniform

Based on the results presented in Figs. 6.10 and 6.11, it can be seen that the result for a specific value a is identical to that for $a + 90°$ by rotating the tunnel axis by 90° as long as the in situ-stresses in the rock mass are uniform.

Fig. 6.12 shows the numerical results of radial stresses in the space surrounding the tunnel after the shotcrete installation for cases where the in-situ stresses in the rock mass are non-uniform ($k \neq 1$). The distribution of radial stresses exhibits an unsymmetrical pattern to the inclined stratification planes, unless the stratification planes are either horizontal, or vertical. Fig. 6.12 also suggests that the distribution of radial stresses for a specific value a with coefficient k is the same as that for $a + 90°$ with coefficient $1/k$, by rotating the tunnel axis by $90°$.

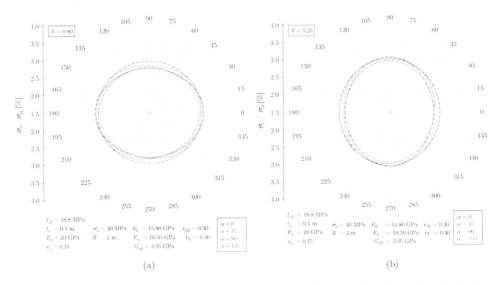

Fig. 6.13. Distributions of Radial Stresses along the Rock-Shotcrete Interface when the In-Situ Stresses in the Rock Mass are Non-Uniform

The distribution of radial stresses along the rock-shotcrete interface as a result of the installation of a 10 cm shotcrete is presented in Fig. 6.13. For $k = 0.80$ (Fig. 6.13a) and if $a = 0°$, the maximum scaled radial stress, σ_r/σ_o, equals to 3.08% was found at the sidewalls, whereas the minimum scaled radial stress equals to 2.75% was observed at the roof and invert. If $a = 90°$, the maximum scaled radial stress equals to 3.05% was located at the sidewalls, whereas the minimum scaled radial stress equals to 2.97% was found at the roof and invert. If $a = 45°$, the maximum scaled radial stress equals to 3.02% was obtained at 15° and 195° measured counterclockwise from the x-axis, the minimum scaled radial stress equals to 2.80% was situated at 105° and 285°. If $a = 135°$, the maximum scaled radial stress equals to 3.02% was found at 165° and 345°, whereas the minimum scaled radial stress equals to 2.80% was observed at 75° and 255°.

Analogously, for $k = 1.25$ (Fig. 6.13b) and when $a = 0°$, the maximum scaled radial stress equals to 3.05% was found at the roof and invert, while the minimum scaled radial stress equals to 2.97% was observed at the sidewalls. When $a = 90°$, the maximum scaled radial stress equals to 3.08% was obtained at the roof and invert, while the minimum scaled radial stress equals to 2.78% was located at the sidewalls. When $a = 45°$, the maximum scaled radial stress equals to 3.02% was found at 75° and 255°, whereas the minimum scaled radial stress equals to 2.80% was found at

165° and 345°. When $a = 135°$, the maximum radial stress equals to 3.02% was found at 105° and 285°, whereas the minimum scaled radial stress equals to 2.80% was observed at 15° and 195°. The numerical results of compressive radial stresses along the rock-shotcrete interface obtained for $k = 0.80$, $k = 1.00$ and $k = 1.25$ are summarized in Table 6.4. Since the stresses in the shotcrete are still below the compressive strength of concrete, f_{ck}, type C20/25, a 10-cm shotcrete is acceptable and can be used as a support system to reduce deformations in the rock mass.

Table 6.4. Radial Stresses along the Rock-Shotcrete Interface

	σ_r (MPa)											
ϑ	$k = 0.80$				$k = 1.00$				$k = 1.25$			
	a				a				a			
	0°	45°	90°	135°	0°	45°	90°	135°	0°	45°	90°	135°
0°	1.23	1.19	1.22	1.19	2.98	2.90	2.86	2.90	1.19	1.13	1.11	1.13
15°	1.22	1.20	1.22	1.18	2.96	2.87	2.86	2.93	1.19	1.14	1.12	1.12
30°	1.19	1.20	1.20	1.16	2.93	2.86	2.87	2.96	1.19	1.17	1.13	1.13
45°	1.16	1.19	1.20	1.14	2.90	2.86	2.90	2.98	1.20	1.19	1.16	1.14
60°	1.13	1.17	1.19	1.13	2.87	2.86	2.93	2.96	1.20	1.20	1.19	1.16
75°	1.12	1.14	1.19	1.12	2.86	2.87	2.96	2.93	1.21	1.21	1.22	1.18
90°	1.11	1.13	1.19	1.13	2.86	2.90	2.98	2.90	1.22	1.19	1.23	1.19
105°	1.12	1.12	1.19	1.14	2.86	2.93	2.96	2.87	1.21	1.18	1.22	1.21
120°	1.13	1.13	1.19	1.17	2.87	2.96	2.93	2.86	1.20	1.16	1.19	1.20
135°	1.16	1.14	1.20	1.19	2.90	2.98	2.90	2.86	1.20	1.14	1.16	1.19
150°	1.19	1.16	1.20	1.20	2.93	2.96	2.87	2.86	1.19	1.13	1.13	1.17
165°	1.22	1.18	1.22	1.21	2.96	2.93	2.86	2.87	1.19	1.12	1.12	1.14
180°	1.23	1.19	1.22	1.19	2.98	2.90	2.86	2.90	1.19	1.13	1.11	1.13
195°	1.22	1.20	1.22	1.18	2.96	2.87	2.86	2.93	1.19	1.14	1.12	1.12
210°	1.19	1.20	1.20	1.16	2.93	2.86	2.87	2.96	1.19	1.17	1.13	1.13
225°	1.16	1.19	1.20	1.14	2.90	2.86	2.90	2.98	1.20	1.19	1.16	1.14
240°	1.13	1.17	1.19	1.13	2.87	2.86	2.93	2.96	1.20	1.20	1.19	1.16
255°	1.12	1.14	1.19	1.12	2.86	2.87	2.96	2.93	1.21	1.21	1.22	1.18
270°	1.11	1.13	1.19	1.13	2.86	2.90	2.98	2.90	1.22	1.19	1.23	1.19
285°	1.12	1.12	1.19	1.14	2.86	2.93	2.96	2.87	1.21	1.18	1.22	1.21
300°	1.13	1.13	1.19	1.17	2.87	2.96	2.93	2.86	1.20	1.16	1.19	1.20
315°	1.16	1.14	1.20	1.19	2.90	2.98	2.90	2.86	1.20	1.14	1.16	1.19
330°	1.19	1.16	1.20	1.20	2.93	2.96	2.87	2.86	1.19	1.13	1.13	1.17
345°	1.22	1.18	1.22	1.21	2.96	2.93	2.86	2.87	1.19	1.12	1.12	1.14
360°	1.23	1.19	1.22	1.19	2.98	2.90	2.86	2.90	1.19	1.13	1.11	1.13

Fig. 6.14 represents the numerical results of radial deformations along the rock-shotcrete interface for cases where the in-situ stresses in the rock mass are non-uniform ($k \neq 1$). For $k = 0.80$ (Fig. 6.14a) and if $a = 0°$, the radial convergence at the tunnel roof and invert was reduced to 0.29%, while at the sidewalls it was 0.16%. If $a = 45°$, the radial convergence at the tunnel arcs located at 105° and 285° counted counterclockwise from the x-axis, was decreased to 0.27%, while the radial convergence at 15° and 195° was found approximately as 0.18%. If $a = 90°$, the radial convergence at the tunnel roof and invert was found as 0.24%, whereas at the sidewalls it was reduced to 0.21%. If $a = 135°$, the radial convergence at 75° and 255° was decreased to 0.27%, while at 165° and 345° it was calculated as 0.18%.

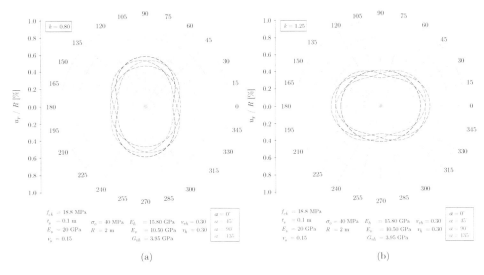

Fig. 6.14. Distributions of Radial Deformations along the Rock-Shotcrete Interface when the In-Situ Stress in the Rock Mass are Non-Uniform

For $k = 1.25$ (Fig. 6.14b) and when $a = 0°$, the radial convergence at the tunnel sidewalls was calculated as 0.24%, while at the tunnel roof and invert it was 0.21%. When $a = 45°$, the radial convergence at 165° and 345° was decreased to 0.27%, whereas at 75° and 255° it was approximately 0.18%. When $a = 90°$, the radial convergence at the sidewalls was found as 0.29% , whereas at the roof or invert it was reduced to 0.16%. When $a = 135°$, the radial convergence located at 15° and 195° was decreased to 0.27%, while at 105° and 285° it was obtained as 0.18%.

6.4. Prestressed Final Lining

The long-term stability of concrete-lined pressure tunnels can be ensured by injecting the circumferential gap between a final lining and shotcrete with a high pressure cement-based grout (Fig. 6.15). As the grout is forced under high pressure, the gap is opened up and filled with densely compacted cement. To allow precise injections in the gap, axial pipes can be embedded along the tunnel walls. As well as debonding agents, synthetic membranes may be placed on the shotcrete surface so as to ease the gap opening.

Since the prestress in the final lining is produced by the support from the surrounding rock mass, this technique is known as the passive prestressing technique. It aims to create adequate prestress in the final lining that offsets tensile stresses induced by the internal water pressure during tunnel operation. It has to be emphasized that the prestress applied to the final lining should be maintained to a level that does not exceed both the smallest principal stress in the rock mass and the compressive strength of concrete, in order to avoid hydraulic jacking of the surrounding rock mass.

Theoretically, as high as 30 bar (3 MPa) of grouting pressure, which is still below the smallest principal stress in the rock mass, can be applied to prestress the final lining. Nevertheless, considering the strain losses due to creep, shrinkage and temperature changes, it was assumed that only 20 bar (2 MPa) of prestress, p_p, remains active at the shotcrete-final lining interface. For the final lining, a 30-cm concrete lining with mechanical properties according to C25/30 (Table 6.3) was used as an example.

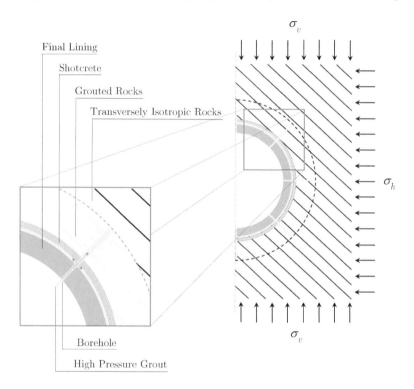

Fig. 6.15. Gap Grouting Procedure

For passive prestressed concrete-lined pressure tunnels, consolidation grouting is a prerequisite that has to be accomplished prior to prestressing a final lining. Since the grout fills and seals discontinuities in the rock mass, the permeability of the grouted rock mass can be reduced, which is favourable in view of limiting seepage into the rock mass. However, since the grout mix cannot penetrate discontinuities in the rock mass with a width greater than 0.1 mm, the permeability of the grouted rock mass cannot be reduced to below 10^{-7} m/s or about 1 Lugeon with cement-based grout (Schleiss, 1986b; Schleiss and Manso, 2012), unless artificial resins such as micro-silica and plasticizers are used. In addition to decreasing the permeability of the rock mass, consolidation grouting can be favourable for tunnel stability since it has more effects on the loosened rock zone (Schleiss, 1986b; Schleiss and Manso, 2012) or at locations where low stresses are dominating (Barton et al., 2001; Vigl and Gerstner, 2009). Therefore, if well grouted, consolidation grouting can potentially reduce the anisotropic rock permeability and deformability.

By using a two-dimensional finite element model, the response of the final lining to prestressing was investigated. In order to reveal the prestress-induced hoop strains along the extrados and intrados of the final lining, the final lining was modelled by using continuum elements. Herein, the same approach as that presented in Simanjuntak et al. (2012b; 2014c) was adopted, i.e. continuous and no slip conditions at the final lining-shotcrete interface.

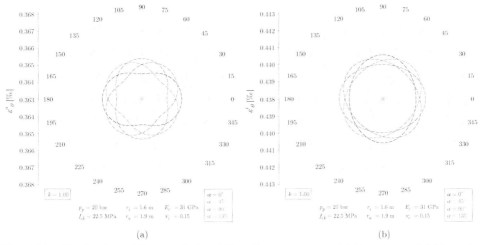

(a)   (b)

Fig. 6.16.   Distributions of Prestress-Induced Hoop Strains along the (a) Extrados, and (b) Intrados of the Final Lining when the In-Situ Stresses in the Rock Mass are Uniform

The gap grouting was modelled by applying uniform compressive load along the shotcrete-final lining interface. To avoid the final lining is influenced by the previous deformations during grouting, it is important to reset the displacements at the shotcrete-final lining interface to zero before simulating the lining prestressing. To reveal stresses in the final lining, the combined Rankine-Von Mises yield criteria (Feenstra, 1993) was used. The former describes the tensile regime, whereas the latter describes the compressive regime.

The numerical results of prestress-induced hoop strains in the final lining for cases where the in-situ stresses are uniform ($k = 1$) are presented in Fig. 6.16. Whereas Fig. 6.16a shows the results obtained along the extrados of the final lining, Fig. 6.16b depicts the results obtained along its intrados. A slight degree of compressive hoop strains was induced throughout the final lining as a result of prestressing works. In the absence of internal water pressure, the compressive hoop strains along the intrados of the final lining are greater than those along the extrados, which is in accordance with the thick-walled cylinder theory (Timoshenko et al., 1970).

By comparing Fig. 6.16a with b, it can be seen that the final lining remains under axisymmetric load. The dip angle, a, governs primarily the distribution of prestress-induced hoop strains in the final lining as long as the in-situ stresses in the rock mass are uniform ($k = 1$). If the stratification planes in the rock mass are horizontal, the final lining is pushed harder against the rock in the horizontal direction than in the vertical direction as the Young's modulus in the horizontal direction is greater than that in the vertical (Fig. 6.16a). As a consequence, the effect of prestressing is more pronounced at the sidewalls of the final lining extrados; inducing high compressive hoop strains at the intrados of the final lining with the maximum strain occurring at the opposite locations, i.e. at the roof and invert of the final lining intrados (Fig. 6.16b). Particular attention should be paid to the strains along the intrados of the final lining, since the maximum internal water pressure is assessed based on these strains.

For $k = 1.00$ (Fig. 6.16b) and if the stratification planes are horizontal, the maximum prestress-induced hoop strain, $\varepsilon^i_{\theta,p_p}$, equals to 0.441‰ was observed at the roof and invert of the final lining intrados, whereas the minimum hoop strain equals to 0.440‰ was found at the sidewalls of the final lining intrados. If $a = 45°$, the maximum prestress-induced hoop strain equals to 0.441‰ was obtained at the arcs of the final lining intrados, specifically at 135° and 315° measured counterclockwise from the x-axis, whereas the minimum hoop strain equals to 0.440‰ was found at 45° and 225°. If the stratification planes are vertical, the maximum prestress-induced hoop strain equals to 0.441‰ was situated at the sidewalls of the final lining intrados, whereas the minimum hoop strain equals to 0.440‰ was found at the roof and invert of the final lining intrados. If $a = 135°$, the maximum prestress-induced hoop strain equals to 0.441‰ was obtained at 45° and 225°, whereas the minimum hoop strain equals to 0.440‰ was found at 135° and 315°. These results imply that as long as the in-situ stresses in the rock mass are uniform, the prestress-induced hoop strain in the lining obtained for a specific value a is identical to that for $a + 90°$ by rotating the tunnel axis by 90°.

In cases where the in-situ stresses in the rock mass are non-uniform ($k \neq 1$), the numerical results are presented in Figs. 6.17 and 6.18, respectively. Whereas Fig. 6.17 displays the prestress-induced hoop strains along the final lining extrados, Fig. 6.18 depicts the prestress-induced hoop strains along the final lining intrados.

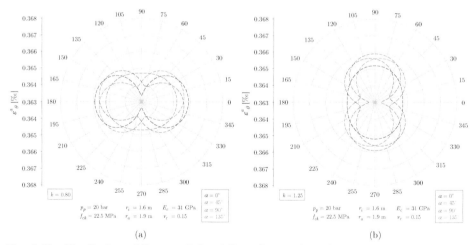

(a)   (b)

Fig. 6.17. Distributions of Prestress-Induced Hoop Strains along the Extrados of the Final Lining when the In-Situ Stresses in the Rock Mass are Non-Uniform

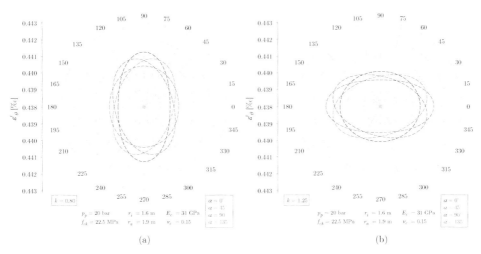

(a) (b)

Fig. 6.18. Distributions of Prestress-Induced Hoop Strains along the Intrados of the Final Lining when the In-Situ Stresses in the Rock Mass are Non-Uniform

Table 6.5. Prestress-Induced Hoop Strains along the Intrados of the Final Lining

ϑ	$\varepsilon'_\theta \times 10^{-4}$											
	$k = 0.80$				$k = 1.00$				$k = 1.25$			
	a				a				a			
	$0°$	$45°$	$90°$	$135°$	$0°$	$45°$	$90°$	$135°$	$0°$	$45°$	$90°$	$135°$
$0°$	4.396	4.398	4.401	4.398	4.400	4.402	4.406	4.402	4.404	4.408	4.413	4.408
$15°$	4.396	4.398	4.401	4.399	4.400	4.401	4.405	4.404	4.404	4.404	4.410	4.409
$30°$	4.398	4.398	4.402	4.402	4.401	4.400	4.404	4.405	4.403	4.401	4.405	4.408
$45°$	4.400	4.399	4.402	4.405	4.402	4.400	4.402	4.406	4.402	4.399	4.400	4.405
$60°$	4.405	4.401	4.403	4.408	4.404	4.400	4.401	4.405	4.402	4.398	4.398	4.402
$75°$	4.410	4.404	4.404	4.409	4.405	4.401	4.400	4.404	4.401	4.398	4.396	4.399
$90°$	4.413	4.408	4.404	4.408	4.406	4.402	4.400	4.402	4.401	4.398	4.396	4.398
$105°$	4.410	4.409	4.404	4.404	4.405	4.404	4.400	4.401	4.401	4.399	4.396	4.398
$120°$	4.405	4.408	4.403	4.401	4.404	4.405	4.401	4.400	4.402	4.402	4.398	4.398
$135°$	4.400	4.405	4.402	4.399	4.402	4.406	4.402	4.400	4.402	4.405	4.400	4.399
$150°$	4.398	4.402	4.402	4.398	4.401	4.405	4.404	4.400	4.403	4.408	4.405	4.401
$165°$	4.396	4.399	4.401	4.398	4.400	4.404	4.405	4.401	4.404	4.409	4.410	4.404
$180°$	4.396	4.398	4.401	4.398	4.400	4.402	4.406	4.402	4.404	4.408	4.413	4.408
$195°$	4.396	4.398	4.401	4.399	4.400	4.401	4.405	4.404	4.404	4.404	4.410	4.409
$210°$	4.398	4.398	4.402	4.402	4.401	4.400	4.404	4.405	4.403	4.401	4.405	4.408
$225°$	4.400	4.399	4.402	4.405	4.402	4.400	4.402	4.406	4.402	4.399	4.400	4.405
$240°$	4.405	4.401	4.403	4.408	4.404	4.400	4.401	4.405	4.402	4.398	4.398	4.402
$255°$	4.410	4.404	4.404	4.409	4.405	4.401	4.400	4.404	4.401	4.398	4.396	4.399
$270°$	4.413	4.408	4.404	4.408	4.406	4.402	4.400	4.402	4.401	4.398	4.396	4.398
$285°$	4.410	4.409	4.404	4.404	4.405	4.404	4.400	4.401	4.401	4.399	4.396	4.398
$300°$	4.405	4.408	4.403	4.401	4.404	4.405	4.401	4.400	4.402	4.402	4.398	4.398
$315°$	4.400	4.405	4.402	4.399	4.402	4.406	4.402	4.400	4.402	4.405	4.400	4.399
$330°$	4.398	4.402	4.402	4.398	4.401	4.405	4.404	4.400	4.403	4.408	4.405	4.401
$345°$	4.396	4.399	4.401	4.398	4.400	4.404	4.405	4.401	4.404	4.409	4.410	4.404
$360°$	4.396	4.398	4.401	4.398	4.400	4.402	4.406	4.402	4.404	4.408	4.413	4.408

Unlike prestressed tunnel linings embedded in transversely isotropic rocks whose stratification planes are either horizontal or vertical and subjected to uniform in-situ stresses, prestressed tunnel linings embedded in transversely isotropic rocks whose stratification planes are inclined and subjected to non-uniform in-situ stresses are no longer under axisymmetric load. Hence, the final lining exhibits non-axisymmetrical deformations even if the applied grouting pressure is uniform. This phenomenon can be seen by comparing Fig. 6.17a with Fig. 6.18a for $k = 0.80$, or Fig. 6.17b with Fig. 6.18b for $k = 1.25$, particularly for the results when $a = 45°$ and $135°$. Consequently, if the in-situ stresses are non-uniform ($k \neq 1$), the distribution of prestress-induced hoop strains in the final lining is determined not only by the dip angle, a, but also by the horizontal-to-vertical stress coefficient, k.

If the stratification planes in the rock mass are either horizontal or vertical, the maximum prestress-induced hoop strain is situated at the roof and invert of the final lining intrados when the in-situ vertical stress is greater than the horizontal. Otherwise, it is located at the sidewalls of the final lining intrados if the in-situ horizontal stress is greater than the vertical. If the stratification planes are inclined, the maximum prestress-induced hoop strain along the final lining intrados is found at the arcs of the final lining intrados (Fig. 6.18).

In the following, the numerical results of prestress-induced hoop strains for cases where the in-situ stresses are non-uniform are analysed. For $k = 0.80$ (Fig. 6.18a) and when the stratification planes are horizontal, the prestress-induced hoop strain at the roof and invert of the final lining intrados was found as 0.441‰, whereas at the sidewalls it was 0.440‰. When the stratification planes are vertical, the prestress-induced hoop strains along the final lining intrados were found as approximately 0.440‰; however, the hoop strain at the roof and invert is slightly higher than that at the sidewalls. When $a = 45°$, the maximum prestress-induced hoop strain equals to 0.441‰ was found at the arcs of the final lining intrados, specifically at 105° and 285° counted counterclockwise from the x-axis, whereas the minimum hoop strain equals to 0.440‰ was observed at 15° and 195°. When $a = 135°$, the maximum prestress-induced hoop strain equals to 0.441‰ was found at 75° and 255°, while the minimum hoop strain equals to 0.440‰ was observed at 165° and 345°.

For $k = 1.25$ (Fig. 6.18b) and if $a = 0°$, the predicted prestress-induced hoop strains along the final lining intrados were found as 0.440‰; nevertheless, the hoop strain at the roof and invert is slightly lower than that at the sidewalls. If $a = 90°$, the prestress-induced hoop strain at the roof and invert of the final lining intrados was found as 0.440‰, while at the sidewalls it was 0.441‰. If $a = 45°$, the maximum prestress-induced hoop strain equals to 0.441‰ was found at 165° and 345°, whereas the minimum hoop strain equals to 0.440‰ was obtained at 75° and 255°. Furthermore, if $a = 135°$, the maximum prestress-induced hoop strain equals to 0.441‰ was found at 15° and 195°, whereas the minimum hoop strain equals to 0.440‰ was observed at 105° and 285°. Again, these results imply that if the in-situ stresses are non-uniform, the distribution of prestress-induced hoop strains in the final lining obtained for a specific value a with coefficient k will be the same as that for $a + 90°$ with coefficient $1/k$, if the tunnel axis is rotated 90°.

It can be concluded that if the in-situ stresses in the rock mass are non-uniform, the effect of prestressing in the final lining is more pronounced at locations where the higher in-situ stress is dominating. If the in-situ vertical stress is greater than the in-situ horizontal stress ($k < 1$), the maximum prestress-induced hoop strain is located at areas around the roof and invert of the final lining intrados (Fig. 6.18a). On the contrary, if the in-situ horizontal stress is greater than the vertical stress ($k > 1$), the maximum prestress-induced hoop strain is situated at areas around the sidewalls of the final lining intrados (Fig. 6.18b). However, since longitudinal cracks will start to develop from the final lining intrados, the assessment of internal water pressure has to be made based on the lowest value of prestress-induced hoop strains at the intrados of the final lining.

6.5. Bearing Capacity of Prestressed Concrete-Lined Pressure Tunnels

When assessing the maximum internal water pressure, it was assumed that the final lining cannot transmit tensile stresses to the rock mass since much of the tensile strength of concrete has already been used in the thermal cooling. To ensure that the hoop stresses in the final lining remain in a compressive state of stress during tunnel operation, the following criterion is used:

$$\varepsilon^i_{\vartheta, p_p} + \varepsilon^i_{\vartheta, p_i} \leq 0 \tag{6.11}$$

where $\varepsilon^i_{\vartheta, p_p}$ and $\varepsilon^i_{\vartheta, p_i}$ are the prestress- and the seepage-induced hoop strain at the final lining intrados, respectively. It must be emphasized that once the internal water pressure is obtained using Eq. (6.11), a certain safety factor has to be applied before putting the value into practice.

Regarding the internal water pressure, two permeable boundaries, which are located at the intrados of the final lining and at the outside of the model domain following the same approach as that presented in Simanjuntak et al.(2012b; 2014c), were introduced in the model. Whereas the permeable boundary at the intrados of the final lining characterizes the hydrostatic head imposed by the internal water pressure, the permeable boundary at the outside of the model domain represents the hydrostatic head due to the groundwater level. The hydrostatic head inside the tunnel was increased until the criterion expressed in Eq. (6.11) is satisfied.

Herein, the permeability coefficients of the rock in the plane of isotropy, k_{rh}, and in the direction normal to the plane of isotropy, k_{rv}, are taken as 10^{-5} m/s and 10^{-6} m/s, respectively. Due to the application of consolidation grouting, it was assumed that the permeability of the rock mass behind the shotcrete is uniform up to a distance of 3 m measured from the tunnel centre. The permeability coefficient of the grouted zone, k_g, the shotcrete, k_s, and the final lining, k_c, can be seen in Table 6.6.

Table 6.6. Permeability Coefficient of Grouted Rock Mass, Shotcrete and Final Lining

k_g (m/s)	k_s (m/s)	k_c (m/s)
10^{-7}	10^{-8}	10^{-9}

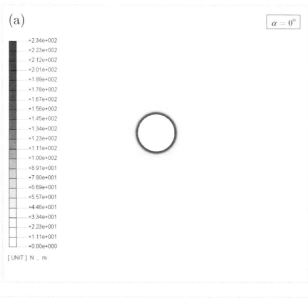

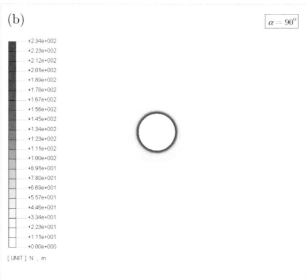

Fig. 6.19. Pore Pressure Head Distributions around the Pressure Tunnel when the Stratifications in the Rock Mass is (a) Horizontal, and (b) Vertical

Based on Eq. (6.11), as high as 234 m of static water head corresponding to a 23 bar (2.3 MPa) of internal water pressure, p_i, was activated. As a result, the hoop strains at the final lining intrados decrease to zero and seepage pressures are generated around the tunnel (Fig. 6.19). Nevertheless, a slight degree of hoop strains in a compressive state of stress (Table 6.7) remained active along the perimeter of the final lining extrados.

Table 6.7. Residual Hoop Strains along the Final Lining Extrados

	$\varepsilon^a_{\vartheta,res} \times 10^{-6}$											
	$k = 0.80$				$k = 1.00$				$k = 1.25$			
ϑ	a				a				a			
	$0°$	$45°$	$90°$	$135°$	$0°$	$45°$	$90°$	$135°$	$0°$	$45°$	$90°$	$135°$
$0°$	5.924	5.916	5.906	5.916	5.910	5.901	5.888	5.901	5.893	5.881	5.860	5.881
$15°$	5.922	5.911	5.905	5.918	5.909	5.895	5.890	5.906	5.894	5.874	5.870	5.893
$30°$	5.917	5.902	5.903	5.917	5.906	5.890	5.895	5.909	5.897	5.879	5.889	5.904
$45°$	5.906	5.890	5.900	5.912	5.901	5.888	5.901	5.910	5.900	5.890	5.906	5.912
$60°$	5.889	5.879	5.897	5.904	5.895	5.890	5.906	5.909	5.903	5.902	5.917	5.917
$75°$	5.870	5.874	5.894	5.893	5.890	5.895	5.909	5.906	5.905	5.911	5.922	5.918
$90°$	5.860	5.881	5.893	5.881	5.888	5.901	5.910	5.901	5.906	5.916	5.924	5.916
$105°$	5.870	5.893	5.894	5.874	5.890	5.906	5.909	5.895	5.905	5.918	5.922	5.911
$120°$	5.889	5.904	5.897	5.879	5.895	5.909	5.906	5.890	5.903	5.917	5.917	5.902
$135°$	5.906	5.912	5.900	5.890	5.901	5.910	5.901	5.888	5.900	5.912	5.906	5.890
$150°$	5.917	5.917	5.903	5.902	5.906	5.909	5.895	5.890	5.897	5.904	5.889	5.879
$165°$	5.922	5.918	5.905	5.911	5.909	5.906	5.890	5.895	5.894	5.893	5.870	5.874
$180°$	5.924	5.916	5.906	5.916	5.910	5.901	5.888	5.901	5.893	5.881	5.860	5.881
$195°$	5.922	5.911	5.905	5.918	5.909	5.895	5.890	5.906	5.894	5.874	5.870	5.893
$210°$	5.917	5.902	5.903	5.917	5.906	5.890	5.895	5.909	5.897	5.879	5.889	5.904
$225°$	5.906	5.890	5.900	5.912	5.901	5.888	5.901	5.910	5.900	5.890	5.906	5.912
$240°$	5.889	5.879	5.897	5.904	5.895	5.890	5.906	5.909	5.903	5.902	5.917	5.917
$255°$	5.870	5.874	5.894	5.893	5.890	5.895	5.909	5.906	5.905	5.911	5.922	5.918
$270°$	5.860	5.881	5.893	5.881	5.888	5.901	5.910	5.901	5.906	5.916	5.924	5.916
$285°$	5.870	5.893	5.894	5.874	5.890	5.906	5.909	5.895	5.905	5.918	5.922	5.911
$300°$	5.889	5.904	5.897	5.879	5.895	5.909	5.906	5.890	5.903	5.917	5.917	5.902
$315°$	5.906	5.912	5.900	5.890	5.901	5.910	5.901	5.888	5.900	5.912	5.906	5.890
$330°$	5.917	5.917	5.903	5.902	5.906	5.909	5.895	5.890	5.897	5.904	5.889	5.879
$345°$	5.922	5.918	5.905	5.911	5.909	5.906	5.890	5.895	5.894	5.893	5.870	5.874
$360°$	5.924	5.916	5.906	5.916	5.910	5.901	5.888	5.901	5.893	5.881	5.860	5.881

In view of pervious concrete, seepage occurs in the rock mass. As a consequence, a bell-shaped saturated zone will develop around the tunnel as the groundwater level is situated below the pressure tunnel (Schleiss, 1997b). Particularly the pore pressure head, their distribution depends on the direction-dependent permeability in the rock mass (Fig. 6.19). Whereas Fig. 6.19a shows the distribution of pore pressure head around the pressure tunnel embedded in transversely isotropic rocks with horizontal stratification planes, Fig. 6.19b illustrates the distribution of pore pressure head around the pressure tunnel embedded in transversely isotropic rocks with vertical stratification planes. Since the rock mass is very permeable compared to the final lining, the shotcrete and the grouted zone, the vertical reach of the seepage flow in the rock mass is small. However, seepage, q, in the order of 0.35 l/s/bar per km length of the tunnel was predicted to develop around the pressure tunnel, which is still below the acceptable value, i.e. 1 l/s/bar per km, as suggested by Schleiss (1988; 2013).

Since the permeability coefficient of the grouted zone is uniform, the seepage pressure at the final lining extrados, p_a, at the shotcrete-grouted zone interface, p_s, and at the extrados of the grouted zone, p_g, for both cases was found as 1.50 bar (0.15 MPa), 0.86 bar (0.09 MPa), and 0.35 bar (0.04 MPa), respectively. While Fig. 6.20 shows the numerical results of residual hoop strains along the final lining extrados, $\varepsilon^a_{\theta,res}$, for cases where the in-situ stresses are uniform, Fig. 6.21 depicts the results for cases where the in-situ stresses are non-uniform. Furthermore, Fig. 6.21a illustrates the results for $k = 0.80$, while Fig. 6.21a presents the results for $k = 1.25$. These results suggest that the redistribution of hoop strains in the final lining after the activation of internal water pressure for a specific value a with coefficient k corresponds to that for $a + 90°$ with coefficient $1/k$, by rotating the tunnel axis by 90°.

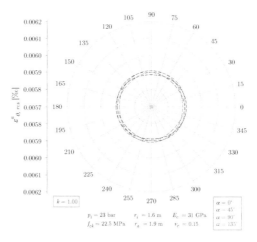

Fig. 6.20. Distributions of Residual Hoop Strains along the Extrados of the Final Lining when the In Situ Stresses in the Rock Mass are Uniform

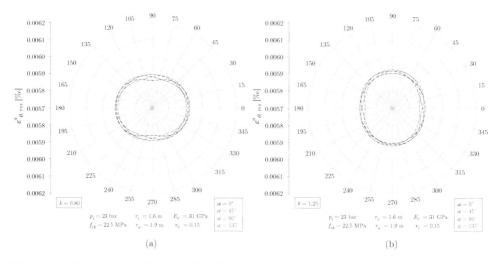

Fig. 6.21. Distributions of Residual Hoop Strains along the Extrados of the Final Lining when the In-Situ Stresses in the Rock Mass are Non-Uniform

As long as the internal water pressure is not greater than 23 bar (2.3 MPa) or when the static water level is not higher than 234 m, the final lining will remain in a compressive state of stress. However, a factor of safety has to be applied to this value so as to account for uncertainties in practice.

In view of the lowest prestress-induced hoop strain at the final lining intrados, it is obvious that as long as the in-situ stresses are uniform, the orientation of stratification planes determines the bearing capacity of the tunnel. This is shown in Fig. 6.16b, since potential locations where longitudinal cracks can occur in the final lining relate to the dip angle.

For cases where the in-situ stresses are non-uniform, the pressure tunnel embedded in transversely isotropic rocks with horizontal stratification planes is the most unfavourable scenario with regard to the bearing capacity if the in-situ vertical stress is greater than the in-situ horizontal stress ($k < 1$) (Fig. 6.18a). Since the lowest hoop strain is located at the sidewalls of the final lining intrados, longitudinal cracks can occur at the sidewalls. On the contrary, if the in-situ horizontal stress is greater than the in-situ vertical ($k > 1$), the tunnel embedded in transversely isotropic rocks with vertical stratification planes is the most unfavourable scenario (Fig. 6.18b) and longitudinal cracks can occur at the roof and invert of the final lining.

Principally, longitudinal cracks will start to develop from the inside of the final lining and more specifically at a location where the tensile strength of concrete is exceeded. A minimum of two cracks can be expected to occur in the final lining for reasons of symmetry. Utilizing information obtained from Figs. 6.16b and 6.18, potential crack locations in the final lining embedded in transversely isotropic rocks can be identified and are summarized in Table 6.8.

Table 6.8. Potential Crack Locations in the Final Lining

k	a	Potential Crack Locations
0.80	0°	Sidewalls, at $\vartheta = 0°$ and 180°
	45°	Arcs, at $\vartheta = 15°$ and 195°
	90°	Sidewalls, at $\vartheta = 0°$ and 180°
	135°	Arcs, at $\vartheta = 165°$ and 345°
1.00	0°	Sidewalls, at $\vartheta = 0°$ and 180°
	45°	Arcs, at $\vartheta = 45°$ and 225°
	90°	Roof and Invert, at $\vartheta = 90°$ and 270°
	135°	Arcs, at $\vartheta = 135°$ and 315°
1.25	0°	Roof and Invert, at $\vartheta = 90°$ and 270°
	45°	Arcs, at $\vartheta = 75°$ and 255°
	90°	Roof and Invert, at $\vartheta = 90°$ and 270°
	135°	Arcs, at $\vartheta = 105°$ and 285°

Finally, once the final lining is cracked, high local seepage can take place around the crack opening. In such circumstances, using an overall rock mass permeability equals to the highest permeability that the rock may have, a simple approach introduced by Simanjuntak et al. (2013) can be applied as a rough estimation of seepage associated with cracks around prestressed concrete-lined pressure tunnels embedded in transversely isotropic rocks.

6.6. Concluding Remarks

By means of a two-dimensional finite element model code DIANA, the behaviour of prestressed concrete-lined pressure tunnels embedded in elastic transversely isotropic rocks subjected to either uniform or non-uniform in-situ stresses was investigated. Two distinctive cases were examined based on whether the in-situ vertical stress is higher, or lower than the in-situ horizontal stress.

As long as the in-situ stresses are uniform, the distribution of stresses and deformations around the tunnel will demonstrate a symmetrical pattern to the orientation of stratification planes, a. In such cases, the distribution of stresses and deformations for a specific value a is the same as that for $a + 90°$, by rotating the tunnel axis by 90°. However, this is not the case when the in-situ stresses are non-uniform. As well as the dip angle, the in-situ stress ratio, k, affects the load sharing between the rock mass and the final lining. Whereas horizontal and vertical stratification planes still contribute to a symmetrical pattern of the distribution of stresses and deformations around the tunnel, inclined stratification planes contribute an unsymmetrical pattern. Thereby, the distribution of stresses and deformations for a specific value a with coefficient k is identical to that for $a + 90°$ with coefficient $1/k$ by rotating the tunnel axis by 90°. Yet, the saturated zone in the rock mass as a result of seepage is exclusively governed by the orientation of stratification planes in the rock mass.

The superposition principle, which is the sum of prestress- and seepage-induced hoop strains at the final lining intrados, was again used in this research to determine the bearing capacity of prestressed concrete-lined pressure tunnels embedded in transversely isotropic rocks. As well as to assessing the maximum internal water pressure, this criterion applies to identifying potential crack locations in the final lining once the hoop strains in the final lining intrados during tunnel operation exceeds the proof tensile strain of concrete. This research suggests that if the in-situ vertical stress is equal to the in-situ horizontal stress, locations of longitudinal cracks in the final lining relate to the orientation of stratification planes. If the in-situ vertical stress is greater than the in-situ horizontal stress and the stratification planes in the rock mass are horizontal, longitudinal cracks are likely to occur at the sidewalls of the final lining. If the in-situ horizontal stress is greater than the in-situ vertical stress and the stratification planes are vertical, longitudinal cracks can occur at the roof and invert of the final lining. Furthermore, if the in-situ vertical stress and the in-situ horizontal stress are unequal and the stratification planes in the rock mass are inclined, longitudinal cracks can occur at the arcs of the final lining and their locations are influenced by the orientation of stratification planes and the in-situ stress ratio.

It is however worth mentioning that this research was carried out based on the main assumption that the rock mass supporting the tunnel was considered as an elastic transversely isotropic material with one direction of the plane of transverse isotropy. If the behaviour of the rock mass covering the tunnel is controlled by persistent discontinuities that ultimately control the rock behaviour, or the plane of transverse isotropy does not strike parallel to the tunnel axis, the load sharing between the rock mass and the final lining needs to be investigated using an approach going beyond the ones introduced herein so as to acquire more accurate results.

7 | Longitudinal Cracks in Pressure Tunnel Concrete Linings[6,7]

Concrete-lined pressure tunnels are subjected to high internal water pressure during operation. When the hoop stress at the final lining intrados exceeds the tensile strength of concrete, longitudinal cracks can occur. As a result of lining cracking, high local seepage can take place around the crack openings. Seepage at high pressure can induce the washing out of the joint fillings in the rock mass and endanger tunnel stability.

Whether or not the internal water pressure acts entirely at the final lining extrados, depends primarily on the crack widths in the final lining. The width of cracks can be estimated based on the total circumferential deformations of the surrounding rock mass, which is governed not only by mechanical boundary pressures, but also by seepage pressures. In turn, seepage around the tunnel depends not only on the permeability of concrete, grouted zone and rock mass, but also on the width of cracks. The latter is mainly responsible for the quantity of seepage around the tunnel. Thus, assessing seepage and seepage pressures associated with longitudinal cracks requires solutions dealing with this coupling behaviour.

This chapter deals with one specific problem: cracking in pressure tunnel linings. A simplified approach to assess the seepage as well as seepage pressure associated with longitudinal cracks is introduced. However, numerical models are needed to capture the propagation of longitudinal cracks in the lining and the overall distribution of seepage around a cracked tunnel. The relevance and benefits of analytical and numerical solutions are outlined.

---

[6] Based on Simanjuntak, T.D.Y.F., Marence, M., Mynett, A.E., Schleiss, A.J. (2013). *Mechanical-Hydraulic Interaction in the Cracking Process of Pressure Tunnel Linings.* Hydropower & Dams, 20(5): 112-119.

[7] Based on Simanjuntak, T.D.Y.F., Marence, M., Mynett, A.E., Schleiss, A.J. (2014). *Longitudinal Cracks in Pressure Tunnels: Numerical Modelling and Structural Behaviour of Passive Prestressed Concrete Linings.* In: Numerical Methods in Geotechnical Engineering. CRC Press, pp. 871-875. ISBN 978-1-138-00146-6.

7.1. Introduction

As soon as the hoop strain at the final lining intrados exceeds the proof tensile strain of concrete, longitudinal cracks will occur. The longitudinal cracks, which will take place at locations submitted to the smallest total stress (Simanjuntak et al., 2014c), can induce high local seepage around the crack openings. Weak zones, such as tunnel roof and transition floor-wall are also vulnerable to longitudinal cracks (Schleiss, 1986b) because tensile strength of concrete cannot entirely be preserved throughout. Therefore, if the passive prestressing technique is used, it is necessary to maintain the final lining in a compressive state of stress during tunnel operation in order to avoid high seepage into the rock mass.

Together with high pressure, seepage through crack openings can induce the washing out of joint fillings in the rock mass. If left untreated, a landslide can occur when pressure tunnels are situated close to valley slopes. This erosion process can cause hydraulic jacking or hydraulic fracturing of the rock mass. Whereas the hydraulic jacking is the opening of existing cracks or joints in the rock mass due to high internal water pressure, the hydraulic fracturing is the event that produces fractures in a sound rock. As a conservative rule, one should consider that cracks will or can exist in any rock mass so that only the case of the opening of existing discontinuities may in fact be of concern in the practice.

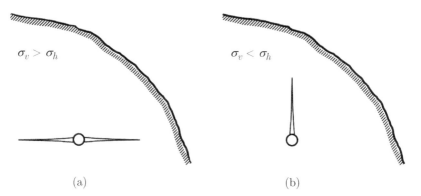

$\sigma_v > \sigma_h$

$\sigma_v < \sigma_h$

(a)

(b)

Fig. 7.1. Schematic Possible Tunnel Failures

A number of tunnel failures and accidents have been encountered in the past as a result of severe cracking of concrete linings (Broch, 1982; Deere, 1983). Some of these are associated with either hydraulic jacking or hydraulic fracturing, and have been the cause of extensive repairs and huge loss in energy production. A proper approach to estimate seepage and seepage pressures as a result of cracking of pressure tunnel linings is therefore of great importance, not only for assessing seepage into the rock mass, but also for taking appropriate measures in view of tunnel safety.

This chapter deals with the cracking of pressure tunnel linings. It aims at quantifying seepage around the pressure tunnel as a result of crack openings in the final lining. In the analysis, three zones are considered, namely the final lining, the grouted zone and the rock mass (Fig. 7.2).

In this chapter, the rock mass is assumed to behave as an elastic isotropic material. The modified load-line diagram method (Simanjuntak et al., 2013) is adequate to calculate the prestress-induced hoop strain in the final lining. The concept to determine the internal water pressure resulting in longitudinal cracks in the lining is oriented towards the maximum utilization of the tensile strength of concrete. A step-by-step calculation procedure is introduced to reveal the two unknowns, i.e. the seepage pressure behind the crack openings and the total seepage out of a cracked concrete-lined pressure tunnel. However, it is necessary by means of numerical models to predict the saturated zone around a cracked tunnel. Besides a simple method to quickly assess the seepage and seepage pressures associated with longitudinal cracks, this will be the second innovative aspect of this research.

7.2. Cracking in Pressure Tunnel Linings

As long as longitudinal cracks in the final lining can be avoided, a criterion to assess the maximum internal water pressure given by Simanjuntak et al. (2012a; 2012b; 2014c) is adequate. However, once longitudinal cracks occur in the final lining, the determination of seepage into the rock mass has to take into account for the effects of crack openings.

If not sealed with waterproofing measures, a concrete lining will permit seepage into the rock mass. The seepage may not be detrimental as long as the seepage pressure in the rock mass is low and no washing out of the joint fillings in the rock mass would occur. But if the internal pressure is high enough to open the cracks in the final lining, seepage pressure behind the lining will increase. This can generate higher seepage pressure into the rock mass, which can endanger tunnel stability (Fig. 7.2b).

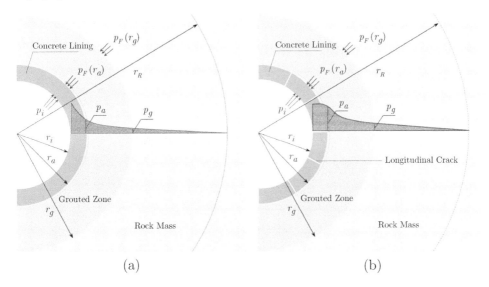

Fig. 7.2. Schematic Development of Seepage Pressures (a) Before, and (b) After Lining Cracking

7.3. Basic Principles

Longitudinal cracks will occur in a final concrete lining as soon as the hoop strain at the final lining intrados exceeds the proof tensile strain of concrete. This criterion can be expressed as (Simanjuntak et al., 2013):

$$\varepsilon^i_{\vartheta, res} = \varepsilon^i_{\vartheta, p_p} + \varepsilon^i_{\vartheta, p_i} \geq \frac{f_{ctk}}{E_c} \tag{7.1}$$

where f_{ctk} represents the design tensile strength of concrete, $\varepsilon^i_{\vartheta, p_p}$ and $\varepsilon^i_{\vartheta, p_i}$ indicate the prestress- and seepage-induced hoop strain at the final lining intrados, respectively.

Without reinforcement, concrete is a brittle material. Due to longitudinal cracks, concrete linings can only transmit radial stresses into the surrounding rock mass. Depending on the degree of crack openings, which is the ratio between the crack width at the extrados and that at the intrados, $2a_a/2a_i$, the radial stresses, $\sigma_r(r_a)$, at the extrados of a cracked final lining whose its magnitude is equal to the mechanical boundary pressure between the lining and the grouted rock, $p_F(r_a)$, can be calculated as (Schleiss, 1986a):

$$\sigma_r(r_a) = p_F(r_a)$$

$$= \frac{(p_a - p_i)}{r_a \left(1 - \dfrac{2a_a}{2a_i}\right)^2 \left(1 + \dfrac{2a_a}{2a_i}\right)} \cdot \tag{7.2}$$

$$\cdot \left[\left(\frac{2a_a}{2a_i}\right)^2 \left(r_a - \left(2 - \frac{2a_a}{2a_i}\right) r_i\right) + \frac{2a_a}{2a_i} r_i + r_a \left(1 - 2\left(\frac{2a_a}{2a_i}\right)\right) \right]$$

The radial deformations at the inner border of the grouted zone, $u_g(r_a)$, can be determined using (Schleiss, 1986a):

$$\frac{u_g(r_a)}{r_a} = \frac{(1 + \nu_g)(p_a - p_g)}{2 E_g (1 - \nu_g)} \cdot \tag{7.3}$$

$$\left[\frac{r_a^2 \left(1 - 2\nu_g + (r_g / r_a)^2\right)}{r_g^2 - r_a^2} + (1 - 2\nu_g)\left(1 + \frac{(1 - \nu_g)}{\ln(r_g / r_a)}\right) \right] -$$

$$- p_F(r_g) \frac{(1 + \nu_g)(1 - 2\nu_g)}{E_g} -$$

$$- \left[p_F(r_g) - \sigma_r(r_a)\right] \frac{r_a^2 (1 + \nu_g)\left(1 - 2\nu_g + (r_g / r_a)^2\right)}{E_g (r_g^2 - r_a^2)}$$

In relation to radial deformations, circumferential deformations at the inner border of the grouted zone can be obtained. The circumferential deformations in the rock mass have to correspond to the total width of longitudinal cracks at the extrados of the final lining.

$$n\,(2a_a) = u_g(r_a)\,2\,\pi \tag{7.4}$$

The crack width at the final lining intrados, $2a_i$, is obtained as (Schleiss, 1986a):

$$(2a_i) = (2a_a)\left(r_a\,/\,r_i\right) \tag{7.5}$$

Based on the compatibility condition of deformations, that is the radial deformation at the outer border of the grouted zone, $u_g(r_g)$, equals to that at the inner border of the rock mass, $u_r(r_g)$, the mechanical boundary pressure at the grouted zone-rock mass interface, $p_F(r_g)$, can be obtained as (Schleiss, 1986b; Simanjuntak et al., 2013):

$$p_F(r_g) = \cfrac{\begin{aligned}&\dfrac{(1+\nu_r)\,E_g}{(1+\nu_g)\,E_r}\,\dfrac{(p_R - p_g)}{2(1-\nu_r)}\cdot\\[2mm]&\cdot\left[\dfrac{r_g^{\,2}}{R^2 - r_g^{\,2}}\left(1 - 2\nu_r + (R\,/\,r_g)^2\right) + (1-2\nu_r)\left(1 + \dfrac{1-\nu_r}{\ln\,(R\,/\,r_g)}\right)\right] -\\[2mm]&- (p_g - p_a)\left[\dfrac{r_a^{\,2}}{r_g^{\,2} - r_a^{\,2}} + \dfrac{(1-2\nu_g)}{2\ln(r_g\,/\,r_a)}\right]\end{aligned}}{\begin{aligned}&\dfrac{(1+\nu_r)\,E_g}{(1+\nu_g)\,E_r}\cdot\\[2mm]&\cdot\left[\dfrac{r_g^{\,2}}{R^2 - r_g^{\,2}}\left(1 - 2\nu_r + (R\,/\,r_g)^2\right)\right] + \left[\dfrac{2\,r_a^{\,2}}{r_g^{\,2} - r_a^{\,2}}\left(1-\nu_g\right) + (1-2\nu_g)\right]\end{aligned}} \tag{7.6}$$

7.4. Seepage Out of Cracked Pressure Tunnels

Seepage around a prestressed concrete-lined pressure tunnel as a result of lining cracking depends not only on the permeability of the lining, the grouted zone and the rock mass, but also mainly on the number of longitudinal cracks as well as the crack width. To investigate whether or not seepage will endanger tunnel stability, one has to consider the mechanical-hydraulic coupling behaviour (Schleiss, 1987; Bian et al., 2009; Simanjuntak et al., 2013). The mechanical-hydraulic coupling behaviour can be described as follows: seepage pressures through longitudinal cracks in the final lining can result in an alteration of rock mass deformations. These deformations influence the crack width in the lining and incipient fractures in the rock mass. As a result, the permeability of the rock mass is changed. The change of rock mass permeability can affect seepage pressures and therefore seepage out of the pressure tunnel.

When the final lining is cracked, seepage developed throughout the pressure tunnel is governed not only by the magnitude of the internal water pressure, but also by the width, $2a$, and the number, n, of cracks in the final lining, the permeability of the final lining between the cracks, k_c, the grouted zone, k_g, and the rock mass, k_r.

The seepage through a cracked concrete lining can be estimated as (Schleiss, 1986a):

$$q_c = \frac{(p_i - p_a) \, 2\pi \, k_c}{\varrho_w \, g \, \ln (r_a \, / \, r_i)} + \frac{2 \, (p_i - p_a) \, (2a_i)^3 \, n \, (2a_a \, / \, 2a_i)^2}{\varrho_w \, (r_a - r_i) \, 12 \, \nu_w \, [1 + (2a_a \, / \, 2a_i)]} \tag{7.7}$$

where p_i is the internal water pressure, p_a is the seepage pressure at the extrados of the final lining, r_i is the internal radius of the final lining, r_a is the external radius of the final lining, ϱ_w is the density of water, g is the gravity acceleration, and ν_w is the kinematic viscosity of water.

Whereas the first components in Eq. (7.7) represent the seepage through a pervious concrete lining, the second components characterize the seepage through the longitudinal cracks.

The seepage through the grouted zone, q_g, can be calculated using:

$$q_g = \frac{(p_a - p_g) \, 2\pi \, k_g}{\varrho_w \, g \, \ln (r_g \, / \, r_a)} \tag{7.8}$$

The seepage pressure at the outer border of the grouted zone, p_g, can be calculated iteratively using (Bouvard, 1975; Simanjuntak et al., 2013):

$$\frac{p_g}{\varrho_w \, g} - \frac{3}{4} r_g = \frac{q_g}{2\pi \, k_r} \ln \frac{q_g}{\pi \, k_r \, r_g} \tag{7.9}$$

Substituting Eq. (7.8) into Eq. (7.9) ensues:

$$q_g = \frac{2\pi \, k_g}{\ln (r_g \, / \, r_a)} \left(\frac{p_a}{\varrho_w \, g} - \frac{q_g}{2\pi \, k_r} \ln \frac{q_g}{\pi \, k_r \, k_g} - \frac{3}{4} r_g \right) \tag{7.10}$$

The seepage through the rock mass, q_r, can be calculated as:

$$q_r = \frac{(p_g - p_R) \, 2\pi \, k_r}{\varrho_w \, g \, \ln (R_v \, / \, r_g)} \tag{7.11}$$

where p_R denotes the seepage pressure in the rock mass, and R_v represents the vertical reach of seepage flow.

The vertical reach, R_v, and the horizontal reach, R_h, of the seepage flow can be calculated as (Schleiss, 1986b):

$$R_v = \left(\frac{q}{\pi \, k_r} \right) \ln (2)$$
$$R_h = \frac{q}{3 \, k_r} \tag{7.12}$$

Furthermore, by applying the continuity condition, that is Eq. (7.7) is equal to Eq. (7.8) and to Eq. (7.11), the seepage pressure at the extrados of the final lining, p_a, can be obtained as (Simanjuntak et al., 2013):

$$p_a = p_i - \cfrac{\dfrac{\pi \, k_r}{\ln \left(R_v \, / \, r_g \right)} \left(p_g - p_R \right)}{\left[\dfrac{\pi \, k_c}{\ln \left(r_a \, / \, r_i \right)} + \dfrac{g \left(2a_i \right)^3 \, n \, \left(2a_a \, / \, 2a_i \right)^2}{12 \, \nu_w \, \left(r_a - r_i \right) \left[\, 1 + \left(2a_a \, / \, 2a_i \right) \right]} \right]} \tag{7.13}$$

7.5. Calculation Procedure

The width of longitudinal cracks at the final lining extrados can be calculated based on the circumferential deformations at the inner border of the grouted zone. These deformations are influenced not only by mechanical boundary pressures, but also by seepage pressures. In turn, seepage pressures generate seepage, which depends not only on the permeability of the lining, the grouted zone and the rock mass, but also on the width of longitudinal cracks.

In the following, a step-by-step calculation procedure is proposed so as to account for the redistribution of seepage pressures as well as seepage around a pressure tunnel, once longitudinal cracks occur in the final lining.

(A) As soon as the prestress-induced hoop strain at the final lining intrados, $\varepsilon^i_{\theta, p_i}$, is obtained, determine the internal water pressure, p_i, and the seepage, q, for the case of uncracked final lining as discussed in Simanjuntak et al. (2012a).

(B) Increase the internal water pressure and investigate the crack status in the final lining according to the condition given by Eq. (7.1). When the tensile strength of concrete is exceeded due to internal water pressure, longitudinal cracks in the final lining occur.

(C) In view of longitudinal cracks, assume seepage pressure acting behind the cracked final lining, p_a, to a value that is close to the internal water pressure resulting in longitudinal cracks.

(D) Assume total seepage, q, in which its magnitude must be higher than that computed in step (A). Recalculate the corresponding seepage pressures at the grouted zone, p_g, and in the rock mass, p_R, using Eqs. (7.8) and (7.11).

(E) Determine the mechanical boundary stress transmitted to the grouted zone, $p_F(r_a)$, and to the rock mass, $p_F(r_g)$, using Eqs. (7.2) and (7.6) respectively, by taking into account Eq. (7.5).

(F) Calculate the radial deformation in the grouted zone, $u_g(r_a)$, using Eq. (7.3). For a given number of cracks, n, the width of cracks at the lining extrados, $2a_a$, and intrados, $2a_i$, can be obtained using Eqs. (7.4) and (7.5).

(G) Once the crack widths at the intrados and extrados of the lining are obtained, calculate the seepage pressure acting at the lining extrados, p_a, using Eq. (7.13) and the corresponding seepage through the cracked lining, q_c, using Eq. (7.7). At the same time, investigate the development of seepage pressure in the grouted zone, p_g, by comparing Eq. (7.8) with Eq. (7.9).

(H) Repeat the calculation step (C) to (G) until the seepage pressures at the final lining extrados, p_a, at the outer border of the grouted zone, p_g, and the seepage around the pressure tunnel, q, remain constant.

7.6. Practical Example

In the following, the proposed calculation procedure is applied in an example so as to calculate seepage pressures as well as seepage around a pressure tunnel occurring before and after the lining cracking. The external radius of the lining is 2.30 m and the lining thickness is 30 cm. The pressure tunnel is embedded in an elastic isotropic rock mass whose in-situ stresses are non-uniform. The mean in-situ stress in the rock mass, σ_o, is 40 MPa, with the in-situ stress ratio, k, of 1.25.

In this example, a 25-bar grouting pressure is applied to enhance the resistance of the final lining against the internal water pressure. The strain losses due to creep and shrinkage are taken as 30%, and the temperature change during watering-up is assumed as $15°$ C. The consolidation grouting is executed up to a radius of 3.30 m.

This example is analogous to that has already been carried out in Chapter 3 where the effective grouting pressure acting on the lining was calculated as 7.1 bar or 0.71 MPa. Nevertheless, to determine the internal water pressure resulting in longitudinal cracks, the concept herein is oriented towards the maximum utilization of the tensile strength of concrete. Parameters used for the analyses are listed in Table 7.1.

Table 7.1. Parameters Used in the Analyses

Material	Symbol	Value	Unit
Rock Mass	E_r	15	GPa
	ν_r	0.25	-
	k_r	10^{-6}	m/s
Grouted Rock Mass	E_q	15	GPa
	ν_q	0.25	-
	k_q	10^{-7}	m/s
Concrete C25/30 (ÖNORM, 2001)	E_c	31	GPa
	ν_c	0.15	-
	f_{cwk}	30	MPa
	f_{ck}	22.5	MPa
	f_{ctm}	2.6	MPa
	f_{ctk}	1.8	MPa
	k_c	10^{-8}	m/s

As a result of prestressing works, the prestress-induced hoop strain at the final lining intrados, $\varepsilon^i_{\theta,\ pp}$, was calculated as 1.91×10^{-4}. The maximum internal water pressure, p_i, resulting in no tensile stresses in the lining was obtained as 17 bar (1.7 MPa). Consequently, seepage in the order of 55.70 l/s per km length of the tunnel can occur around the tunnel. This value is still within the tolerable limit as long as the pressure tunnel is not put at risks.

Table 7.2. Predicted Hoop Stresses, Seepage, Seepage Pressures and Seepage Reach as a Result of Increasing the Internal Water Pressure

Output	p_i (bar)						
	17	18	19	20	21	22	22.5
$\Delta\sigma_\vartheta$ (MPa)	0.00	+0.32	+0.65	+0.98	+1.31	+1.64	+1.80
q (l/s/km)	55.70	58.86	62.01	65.15	68.28	71.40	214.58
p_a (bar)	4.84	5.16	5.47	5.78	6.10	6.42	22.50
p_q (bar)	1.70	1.84	1.97	2.11	2.25	2.39	10.40
R_v (m)	12.3	13.0	13.7	14.4	15.1	15.8	47.4
R_h (m)	18.6	19.6	20.7	21.7	22.8	23.8	71.6
State	-	Tensile	Tensile	Tensile	Tensile	Tensile	Crack

If the internal water pressure applied is greater than 17 bar, the final lining will be in a tensile state of stress. Longitudinal cracks occur in the lining when the internal water pressure assessed using Eq. (7.1) equals to 22.5 bar (2.25 MPa) or when the static water head inside the final lining is 230 m (Table 7.2).

Using Eq. (7.3), the radial deformation at the inner border of the grouted zone was obtained as 0.42 mm, which corresponds to the circumferential deformation of 2.64 mm. In view of geometrical symmetry, at least two cracks exist in the final lining. In this example, longitudinal cracks in the concrete lining are expected to occur at the tunnel roof and invert.

Once the circumferential deformation at the inner border of the grouted zone is obtained, the crack width at the extrados of the final lining can be calculated using Eq. (7.4). With the number of cracks, n, equals to 2, the crack width at the final lining extrados, $2a_a$, was obtained as 1.32 mm, while at the final lining intrados, $2a_i$, it was 1.52 mm. As a result of crack openings in the final lining, the seepage, q, increases to 214.58 l/s per km length of the tunnel, which is equivalent to 9.54 l/s/km/bar.

Because of the crack opening at the final lining extrados, the internal water pressure acts entirely at the extrados of the final lining. As a consequence, the seepage pressure at the outer border of the grouted zone was increased to 10.40 bar (1.04 MPa). A large unacceptable escape of water with high seepage pressure has to be avoided since it can cause not only huge loss of energy production but also hydraulic jacking of the surrounding rock mass.

If no washing out of joint fillings can occur in the rock mass and the smallest in-situ stress in the rock mass is still higher than the internal water pressure behind the lining, taking into account an adequate factor of safety, one of the techniques available to reduce seepage around the pressure tunnel is by grouting the rock mass surrounding the tunnel. In this regard, improving grout quality is more effective than extending the radius of the grouted zone (Simanjuntak et al., 2013).

As illustrated in Fig. 7.3a, rock grouting can be favourable in view of limiting seepage as well as seepage pressure into the rock mass. However, it should be emphasized that reducing the permeability of the rock mass below 10^{-7} m/s or lower than 1 Lugeon with cement-based grout is difficult and can only be achieved with great efforts.

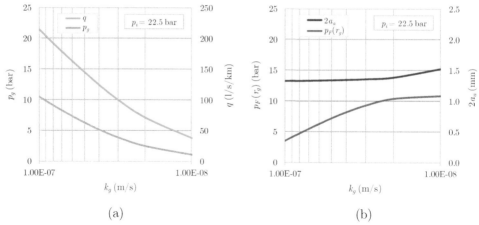

Fig. 7.3. Effect of Grouted Zone Permeability on (a) Seepage and Seepage Pressure in the Grouted Zone, and (b) Crack Width and Pressure Transmitted to the Rock Mass after Cracking of the Final Lining

Furthermore, reducing the permeability of the rock mass close to 10^{-8} m/s by rock grouting is not worthwhile. As illustrated in Fig. 7.3b, the less permeable the grouted zone, the greater the radial deformations in the rock mass and the wider the crack openings will become in the final lining. If the rock mass is too pervious compared to the grouted zone, the crack widths in the final lining increase and they seem to affect the mechanical boundary pressure at the grouted zone-rock mass interface, $p_F(r_g)$.

7.7. Modelling of Cracking of Tunnel Linings

Notably, numerical models can be used to simulate the cracking of pressure tunnel linings. Since concrete specimen in the final lining does not have reinforcement, the concrete behaviour is brittle (Amorim et al., 2014). Consequently, the cracking of prestressed pressure tunnel linings can be modelled based on the discrete cracking approach since the locations of longitudinal cracks can be identified and the number of cracks in the final lining is finite due to geometrical symmetry.

With the discrete cracking approach, a longitudinal crack in the final lining can be modelled as a geometrical discontinuity. In the model, this is done by incorporating interface elements within the original mesh. As soon as hoop stresses along the interface elements exceed the tensile strength of concrete, the node will be split and the discrete crack will be forced to propagate along the element boundary.

If the concrete is assumed to behave in an elastic brittle manner, three parameters namely the normal stiffness modulus, K_n, shear stiffness modulus, K_t, and tensile strength, f_{ctk}, are needed to define the interface elements. Herein, the material properties for the interface used in the simulation, are summarized in Table 7.3.

Table 7.3. Interface Properties

K_n (N/mm^3)	K_t (N/mm^3)	f_{ctk} (MPa)
10^7	10^5	1.8

In order to allow the tunnel to deform in radial direction, the vertical deformations along the horizontal axis and the horizontal deformations along the vertical axis are supported. Herein, the model consists of two consecutive loading. The first loading represents the effective grouting pressure acting at the extrados of the final lining, while the second loading is due to the internal water pressure.

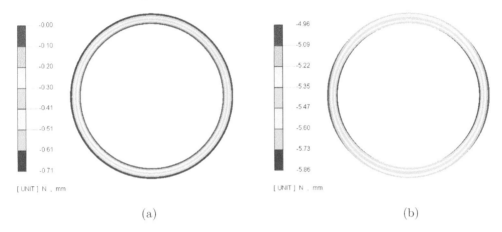

(a)   (b)

Fig. 7.4. Predicted Prestress-Induced (a) Radial, and (b) Hoop Stresses in the Final Lining

The numerical results of prestress-induced radial stresses as well as hoop stresses are presented in Fig. 7.4. The negative sign indicates that the stresses in the lining are in a compressive state. While Fig. 7.4a shows the distribution of radial stresses in the final lining due to the application of grouting pressure, Fig. 7.4b illustrates the distribution of hoop stresses. Particularly the hoop stresses, the final lining is subjected to axisymmetric load when prestressed against the rock mass. The maximum value is located at the sidewalls of the final lining intrados since the in-situ horizontal stress in the rock mass is higher than the in-situ vertical stress. As shown in Fig. 7.4b, the maximum prestress-induced hoop stress was obtained as 5.86 MPa, whereas the minimum hoop stress located at the roof and invert of the final lining was found as 5.73 MPa.

For the second loading, the initial internal pressure was 1 MPa. This pressure was increased until longitudinal cracks occurred in the final lining. The ultimate bearing capacity of the pressure tunnel is reached when the hoop stress at the final lining intrados offsets the tensile strength of concrete. Fig. 7.5 shows the predicted radial and hoop stresses in the final lining before the cracks initiation. While Fig. 7.5a depicts the distribution of radial stresses, Fig. 7.5b illustrates the hoop stresses in the final lining. It can be seen that as much as 1.76 MPa of hoop stresses in a tensile state was induced in the final lining intrados when the internal pressure, was increased up to 1.72 MPa.

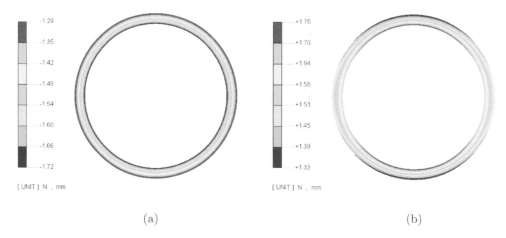

Fig. 7.5. Distribution of (a) Radial, and (b) Hoop Stresses before Lining Cracking

The internal pressure is 17.2 bar (1.72 MPa) is equivalent to the static water head of 175 m. As a result, the maximum tensile hoop stress equals to 1.76 MPa was induced at the roof and invert of the final lining intrados (Fig. 7.5b). If the internal pressure is greater than 1.72 MPa, the final lining is cracked and the cracks will develop at the roof and invert of the final lining intrados (Fig. 7.6).

The numerical results showing the propagation of longitudinal cracks in the final lining are presented in Fig. 7.6. Whereas Fig. 7.6a presents the initial cracks in the final lining, Fig. 7.6d shows the cracking in the final lining at the last numerically stable result. As shown in Fig. 7.6, longitudinal cracks in the final lining are noticed by the separation of nodes along the interface elements. These cracks propagate from the lining intrados towards the lining extrados. Since the cracks start at the final lining intrados, this also implies that the crack width at the final lining intrados is greater than that at the extrados (Schleiss, 1997b; Simanjuntak et al., 2013).

In addition to the cracks propagation, Fig. 7.6 also shows the redistribution of hoop stresses in the final lining. It can be seen that the tensile hoop stresses in the final lining are decreased to zero around the crack openings. At failure, a cracked lining can no longer transmit hoop stresses into the rock mass, but only radial stresses.

Even though the crack propagation in a concrete lining can be simulated using the discrete cracking approach, it has to be acknowledged that the plasticity effects in the concrete are negligible as the behaviour of unreinforced concrete is ideally brittle. Moreover, concrete lining is assumed as an impervious material, rendering that the assessment of the internal water pressure causing longitudinal cracks will be different from that calculated using the simplified analytical approach. Hitherto, for cases of impervious concrete linings, good agreement between analytical and numerical results (Simanjuntak et al., 2014b) as well as between experimental and numerical results (Amorim et al., 2014) is evident.

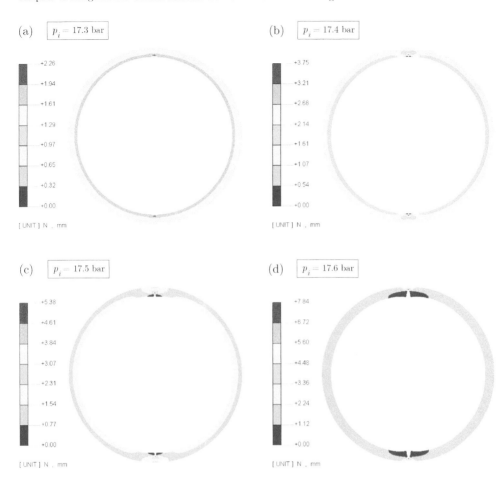

Fig. 7.6. Redistribution of Hoop Stresses after Lining Cracking

Unlike the modelling of cracking in concrete linings, the distribution of pore pressure head around a cracked pressure tunnel can be obtained using DIANA by employing the steady-state groundwater flow analysis independently. To do so, four permeable boundaries can be introduced in the model.

The first permeable boundary is put at the outer model domain that characterizes the groundwater level. The second and the third permeable boundary are located at the locations where longitudinal cracks are expected to occur; in which one permeable boundary is located perpendicular to the roof, and another permeable boundary is located perpendicular to the invert of the final lining intrados. The fourth permeable boundary is applied along the circumferential of the final lining intrados and this represents the internal water pressure.

If the concrete lining is pervious, the simplified analytical approach suggests that the internal water pressure resulting in longitudinal cracks in the lining is obtained as 22.5 bar (2.25 MPa) or when the static water head inside the final lining is 230 m

(Table 7.2). Correspondingly, the numerical results showing the distributions of pore pressure head around the cracked pressure tunnel as a result of the activation of a 230 m static water head are presented in Fig. 7.7. Whereas Fig. 7.7a depicts the distribution of pore pressure head when the longitudinal cracks occur at the lining roof and invert, Fig. 7.7b represents the distribution of pore pressure head when the longitudinal cracks arise at the lining sidewalls.

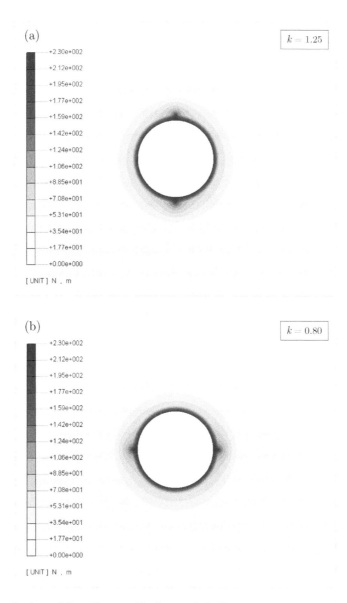

Fig. 7.7. Distributions of Pore Pressure Head around the Pressure Tunnel when Cracks Occurs at the (a) Roof and Invert, and (b) Sidewalls of the Final Lining

As shown in Fig. 7.7, a bell-shaped saturated zone will develop around a pressure tunnels situated above the groundwater level. This result conforms with that has been reported in Schleiss (1986b). However, the numerical results (Table 7.4) suggest that seepage pressures increase at the locations where the cracks occur. An increase in seepage pressures will increase the reach of the seepage flow into the rock mass and eventually affect the saturated zone. Comparing Fig. 7.7a with b, it can be seen that if the longitudinal cracks occur at the lining roof and invert, the vertical reach of seepage flow into the rock mass increases. If the longitudinal cracks occur at the lining sidewalls, the horizontal reach of seepage flow increases. As long as the tunnel safety against hydraulic jacking can be preserved, rock overburden has to ensure that the reach of seepage flow will not detrimentally affect the hydrogeological conditions such as the yield of springs.

Table 7.4. The Vertical Reach of Seepage Flow and Seepage Pressure at the Grouted-Zone and Rock Mass Interface

k	R_v (m)	p_g (bar)	
		Roof and Invert	Sidewalls
1.25	28.9	4.63	4.52
0.80	21.8	4.16	4.99

7.8. Concluding Remarks

A step-by-step calculation procedure to assess seepage and seepage pressures associated with longitudinal cracks in the concrete lining is proposed. This procedure takes into account the mechanical-hydraulic coupling behaviour of pressure tunnels where three zones are considered, namely the final lining, the grouted zone and the ungrouted rock mass. As well as the unknown seepage and seepage pressures behind the cracked lining, the proposed calculation procedure is capable of assessing the internal water pressure resulting longitudinal cracks in the final lining.

The propagation of longitudinal cracks in a concrete lining can be simulated by using numerical models based on the discrete cracking approach. Thereby, a longitudinal crack in the final lining is modelled as a geometrical discontinuity by incorporating interface elements within the original mesh. As a consequence, the numerical results will be dependent on the material properties of the interface elements. Furthermore, using the discrete cracking approach, concrete is assumed as an impervious material. The assessment of the internal water pressure resulting in longitudinal cracks in the lining may provide different results when compared to those calculated by using the assumption of pervious concrete.

Even though the modelling of cracking in pressure tunnel linings is still restricted to particular cases, a realistic distribution of seepage around a cracked concrete-lined pressure tunnel can be obtained using numerical models. Crack openings in the final lining determine the seepage reach into the rock mass and eventually the saturated zone around the pressure tunnel.

When longitudinal cracks occur at the lining roof and invert, the high seepage pressure acting behind the crack openings will extend the vertical reach of seepage flow into the rock mass. On the contrary, when the longitudinal cracks occur at the lining sidewalls, the high seepage pressure behind the crack openings will lengthen the horizontal reach of seepage flow into the rock mass.

As long as the stability of prestressed concrete-lined pressure tunnels against hydraulic jacking can be preserved by adequate rock strength and rock overburden, high seepage into the rock mass can be limited by rock grouting. However, since crack openings are difficult to control with the passive prestressing technique, the washing out of joint fillings in the rock mass can still occur. Therefore, the design criteria for passive prestressed concrete-lined pressure tunnels are: avoiding cracks in a concrete lining, limiting seepage into the rock mass, and ensuring the bearing capacity of the rock mass supporting the tunnel.

8 | Conclusions and Recommendations

8.1. Conclusions

The main objective of this research is to investigate the mechanical and hydraulic behaviour of passive prestressed concrete-lined pressure tunnels. The most important conclusions are summarized herein, followed by recommendations for future research.

8.1.1. A New Design Criterion for Tunnel Bearing Capacity

As long as the rock mass supporting a pressure tunnel is of good quality and the rock overburden is adequate, the bearing capacity of concrete-lined pressure tunnels can be improved by grouting the circumferential gap between the final lining and the rock mass at high pressure. As the grout sets under pressure, a full contact between the final lining and the rock mass is achieved. The load sharing between the rock mass and the final lining can be determined based on the compatibility condition of deformations by putting equal radial deformations at their contact face.

Until now, the bearing capacity of prestressed concrete-lined pressure tunnels is often determined based on the assumption of elastic isotropic rock mass. As a consequence, deformations in the concrete lining are a function of the elastic properties of both the lining and the rock mass. As long as the assumption of elastic isotropic rock mass is acceptable, the load-line diagram method which was developed based on the impervious thick-walled cylinder theory, is still adequate to assess the prestress-induced hoop strains in the lining. However, since concrete is in fact a slightly pervious material, the assessment of the maximum internal water pressure should consider the effects of seepage on rock deformations.

If seepage into the rock mass cannot be avoided, the use of the sole load-line diagram method can result in an overestimation of the maximum internal water pressure (Chapter 3). In this research, a new concept taking into account seepage effects is introduced so as to more appropriately assess the maximum internal water pressure. The maximum internal water pressure can be determined by offsetting the seepage-induced hoop strain at the final lining intrados against the prestress-induced hoop strain. Moreover, a certain factor of safety has to be applied to the predicted value before putting it into practice. In Austria, a safety factor can vary between 1.35 and 1.50 with reference to ÖNORM (2001).

If the passive prestressing technique is used to enhance the tunnel bearing capacity, the grouting pressure injected into the circumferential gap should remain below the smallest principal stress of rock mass so as to avoid hydraulic jacking. Consolidation grouting is also a prerequisite before prestressing the final lining. This is meant not only to homogenize the stress pattern around the tunnel but also to limit seepage, which is favourable for tunnel tightness and stability.

8.1.2. Behaviour of Pressure Tunnels in Isotropic Rock Masses

In cases where there is no preferred orientation of joints within the rock mass, the rock mass can be treated as an isotropic material. Nevertheless, the rock mass may no longer deform elastically due to tunnel excavation. A plastic zone can develop in the rock mass rendering the determination of bearing capacity of prestressed concrete-lined pressure tunnels embedded in elasto-plastic isotropic rock mass requires a method going beyond that under the assumption of elastic isotropic rock mass.

Because the rock strength depends on the in-situ stresses in a non-linear manner, the Hoek-Brown failure criterion can be used to investigate the response of elasto-plastic isotropic rock mass to tunnelling. For prestressed concrete-lined pressure tunnels, the tunnelling construction process includes underground excavation, installation of support, installation of final lining and gap grouting.

Tunnel excavation and support installation involve three-dimensional tunnel advance and pre-relaxation ahead of the tunnel face. As long as the deformations at the support-rock mass interface are known, the load transferred to the support as a result of the tunnel excavation can be obtained. When using two-dimensional models, one of the modelling techniques is by reducing the modulus of elasticity of the rock mass being excavated referred to as the stiffness reduction method.

In this research, the deformation at the support-rock mass interface was assessed by using the convergence-confinement method. In the model, the modulus of elasticity of the rock mass being excavated can be reduced to a certain value so that the numerical results of radial deformations at the support-rock mass interface fit those calculated by using the convergence-confinement method (Chapter 4). As well as the assessment of the load transferred to support, the convergence-confinement method can provide the appropriate location of the support installation with respect to elasto-plastic behaviour of the rock mass. However, this solution is applicable when the tunnel geometry is circular and the in-situ stresses in the rock mass are uniform.

If the in-situ stresses in the rock mass are non-uniform but as long as one of its components acts parallel to the longitudinal axis of excavation, the load transferred to the support can be assessed based on simultaneous tunnel excavation and support installation (Chapter 5 and 6). Since the three-dimensional problem of excavation is not considered, this approach may result in an overestimation of stresses in the support. However, the effects of large deformations on the final lining can be avoided during prestressing by resetting the deformation at the support-final lining interface to zero before simulating the gap grouting.

As soon as the equilibrium condition of tunnel excavation and support installation is reached, the final lining can be installed on the support, such as shotcrete. Together with the rock mass, it is responsible to withstand the load imposed by the internal water pressure. To keep the final lining free from tensile stresses during tunnel operation, the final lining can be prestressed using the passive prestressing technique. In practice, the prestressing of final lining can be executed in form of injections around the tunnel.

By pumping cement-based grout at high pressure into the contact joint between the shotcrete and the final lining, a circumferential gap between the final lining and the shotcrete is opened up and filled with grout. Besides a certain prestress level in the lining, a full contact in the system is provided as the grout hardens.

Taking into account strain losses due to creep, shrinkage and temperature effects, the effective grouting pressure can be assessed using the load-line diagram method. The prestressing of final lining can be modelled by applying a uniform compressive load along the shotcrete-final lining interface. The combined Rankine-Von Mises yield criteria is used to reveal stresses in the shotcrete and the final lining. While the former describes the tensile regime, the latter controls the compressive regime.

For cases when the assumption of elasto-plastic isotropic rock mass is acceptable, the behaviour of pressure tunnels was explored based on whether the in-situ stresses are uniform (Chapter 4) or non-uniform (Chapter 5). The degree of the in-situ stress non-uniformity can be expressed by an in-situ stress ratio, k. If the in-situ stress ratio is one, the in-situ stresses in the rock mass are uniform. Otherwise, two typical cases of non-uniform in-situ stresses are distinguished. If the in-situ stress ratio is less than one, the in-situ vertical stress is greater than the horizontal. If the in-situ stress ratio is more than one, the in-situ horizontal stress is greater than the vertical.

Chapters 4 and 5 showed that the distribution of load sharing between the rock mass and the final lining is governed by the in-situ stress in the rock mass. Unlike for cases where the in-situ stresses are uniform, the load sharing between the rock mass and the final lining for cases when the in-situ stresses are non-uniform is non-uniformly distributed. The distribution of stresses and deformations for a specific value of k, is similar to the case with coefficient $1/k$ by rotating the tunnel axis by 90°.

The maximum internal water pressure for pressure tunnels embedded in an isotropic rock mass subjected to either uniform or non-uniform in-situ stresses is determined by offsetting the seepage-induced hoop strains at the final lining intrados against the prestress-induced hoop strains. Regarding seepage calculations, two permeable boundaries located at the final lining intrados and at the outside of the model domain can be introduced to the model. While the permeable boundary at the final lining intrados represents the hydrostatic head due to the internal water pressure, the permeable boundary at the outside of the model domain corresponds to the hydrostatic head imposed by the groundwater level. Using the steady state groundwater flow analysis, pore pressure head around the tunnel can be generated.

If the maximum internal water pressure is high enough to induce tensile stresses in the final lining, locations where longitudinal cracks can occur in the final lining embedded in elasto-plastic isotropic rock mass whose in-situ stresses are non-uniform can be identified. When the in-situ vertical stress in the rock mass is greater than the in-situ horizontal stress, the longitudinal cracks will occur at the sidewalls of the final lining intrados. Conversely, when the in-situ horizontal stress is greater than the in-situ vertical stress, cracks will occur at the roof or invert of the final lining intrados.

Furthermore, if the smallest in-situ stress is too low to withstand the internal water pressure acting behind the final lining, hydraulic jacking or hydraulic fracturing can occur. While the hydraulic jacking is the opening of existing cracks or joints in the rock mass, the hydraulic fracturing is the event that produces fractures in a sound rock. In order to avoid such failures, it is essential to avoid tensile stresses in the final lining during tunnel operation. A certain factor of safety has to be applied to the maximum internal water pressure before putting it into practice.

8.1.3. Behaviour of Pressure Tunnels in Anisotropic Rocks

Pressure tunnels may be built in an inherently anisotropic rock mass, such as metamorphic rocks. These types of rocks, which are composed of laminations of intact rocks, can take the form of transverse isotropy commonly configured by one direction of planes perpendicular to the direction of deposition. In such formations, the rock mass supporting a pressure tunnel may exhibit significant distinctive deformability and permeability in the direction parallel and perpendicular to the stratification planes. The behaviour of pressure tunnels embedded in transversely isotropic rocks can deviate from that investigated under the assumption of isotropic rocks.

In Chapter 6, an important aspect that is frequently ignored in the design of pressure tunnels, namely the interplay between the dip angle, a, and the in-situ stress ratio, k, was discussed. It was investigated how these two issues affect the lining performance. Providing that there is no slip between the planes of transverse isotropy, an anisotropic rock mass as a whole can be idealized as an elastic transversely isotropic material. As long as the strike of the stratification planes in the rock mass is parallel to the tunnel axis, the mechanical and hydraulic behaviour of pressure tunnels can be investigated by means of a two-dimensional finite element model.

The response of transversely isotropic rocks to a circular excavation can be predicted by means of the elasto-plastic Jointed Rock model. Moreover, the elastic behaviour of the rock mass can be ensured by providing an adequate cohesion along the sliding planes. Regarding the modelling of support installation, prestressing of final lining and activation of internal water pressure, the modelling approach is similar to that for cases where pressure tunnels embedded in elasto-plastic isotropic rock mass subjected to non-uniform in-situ stresses (Chapter 5). The load sharing between the rock mass and the final lining is explored based on whether the in-situ stresses in the rock mass are uniform or non-uniform.

For cases of pressure tunnels embedded in transversely isotropic rocks and if the in-situ stresses are uniform, the final lining will remain under axisymmetric loads. The distribution of load sharing between the rock mass and the final lining demonstrates a symmetrical pattern to the orientation of stratification planes. As long as the in-situ stresses are uniform, the distribution of stresses and deformations for a specific value a is the same as that for $a + 90°$, by rotating the tunnel axis by $90°$.

If the in situ-stresses are non-uniform and the orientation of stratification planes are either vertical or horizontal, the distribution of load sharing between the rock mass and the final lining also demonstrates a symmetrical pattern. Nevertheless, this is not the case when the stratification planes are inclined. In such cases, the final lining is no longer under axisymmetric loads. Even though the distribution of load sharing between the rock mass and the final lining still demonstrates a symmetrical pattern, its major and minor axis are neither parallel nor perpendicular to the orientation of stratification planes (Chapter 6). If the in-situ stresses are non-uniform, this research suggests that the distribution of stresses and deformations for a specific value a with coefficient k is identical to that for $a + 90°$ with coefficient $1/k$ by rotating the tunnel axis by $90°$.

By offsetting the seepage-induced hoop strains at the final lining intrados against the prestress-induced hoop strains, the maximum internal water pressure can be assessed. This approach is also capable of identifying potential crack locations in the final lining embedded in transversely isotropic rocks once the hoop stresses in the final lining exceed the tensile strength of concrete.

If the in-situ vertical stresses are uniform, locations of longitudinal cracks in the final lining are mainly influenced by the orientation of stratification planes. If the in-situ vertical stress is greater than the in-situ horizontal and the stratification planes in the rock mass are horizontal, longitudinal cracks can occur at the sidewalls of the final lining. If the in-situ horizontal stress is greater than the in-situ vertical and the stratification planes are vertical, longitudinal cracks can occur at the roof and invert of the final lining. If the in-situ vertical stress and the in-situ horizontal stress are non-uniform and the stratification planes in the rock mass are inclined, longitudinal cracks can take place at the arcs of the final lining and their locations are influenced by both the orientation of stratification plane and the in-situ stress ratio. Yet, the saturated zone around a pressure tunnel embedded in transversely isotropic rocks is exclusively influenced by the orientation of stratification planes.

8.1.4. Cracking in Pressure Tunnel Lining

As soon as the hoop stress at the final lining intrados exceeds the tensile strength of concrete, longitudinal cracks occur in the final lining. As a result of crack openings, the internal water pressure will act at the final lining extrados and cause high local seepage. If left untreated, seepage can induce the washing out of joint fillings in the rock mass and increase the risk of hydraulic jacking of the surrounding rock mass. If pressure tunnels are situated close to valley slopes with quite impermeable rock cover, even small seepage can ultimately cause a landslide.

The effectiveness of the internal water pressure at the final lining extrados depends predominantly on the number and the width of cracks. The crack width can be calculated based on the circumferential deformation of the rock mass, which is governed not only by mechanical boundary pressures, but also by seepage pressures. In turn, seepage pressures generate seepage from the tunnel, in which its magnitude depends not only on the permeability of the rock mass, the grouted zone, and the concrete lining, but also on the width of longitudinal cracks. Assessing seepage and seepage pressures associated with longitudinal cracks requires solutions dealing with this coupling behaviour.

A step-by-step calculation procedure taking into account the lining cracking history was proposed in Chapter 7 to quantify seepage and thus seepage pressures associated with longitudinal cracks around a pressure tunnel. The approach is developed based on the assumption of elastic isotropic rock mass. Regarding the assessment of internal water pressure resulting in longitudinal cracks, the concept is oriented towards the maximum utilization of the tensile strength of concrete. Once the final lining is cracked, an abrupt increase of seepage will take place around the crack openings. To assess the saturated zone around a pressure tunnel after lining cracking, numerical models can be used.

In cases of passive prestressed concrete-lined pressure tunnels, concrete is ideally a brittle material. Consequently, the propagation of longitudinal cracks in the lining can be simulated based on the discrete cracking concept by incorporating interface elements within the original mesh. Nevertheless, it has to be acknowledged that this concept assumes concrete as an impervious material. This can result in an underestimation of the maximum internal water pressure when compared to that calculated using the proposed calculation procedure with the assumption of pervious concrete.

As a result of seepage, saturated zone around a cracked concrete-lined pressure tunnel can be predicted by performing the steady-state groundwater flow analysis. In the model, permeable boundaries can be introduced at the interface where longitudinal cracks are expected to occur. Seepage pressure through permeable boundaries determines the reach of seepage flow and eventually the saturated zone. If longitudinal cracks occur at the lining roof and invert, high seepage pressure behind the crack openings will extend the vertical reach of seepage flow into the rock mass. On the contrary, if longitudinal cracks occur at the lining sidewalls, high seepage pressure will lengthen the horizontal reach of seepage flow.

Looking at the large scale in the vicinity of pressure tunnels, it is necessary to consider that cracks exist in any rock mass. Only if the stability of passive prestressed concrete-lined pressure tunnels against hydraulic jacking or fracturing can be guaranteed provided by adequate rock strength or rock overburden, seepage into the rock mass can still be tolerated. Otherwise, no longitudinal cracks are allowed to occur in the final lining since crack openings that can induce the washing out of joint fillings in the rock mass are difficult to control with the passive prestressing technique. The design criteria for passive prestressed concrete-lined pressure tunnels are therefore: avoiding cracks in the final lining, limiting seepage into the rock mass, and ensuring the bearing capacity of the rock mass supporting the tunnel.

8.2. Recommendations

This research contributes to a better understanding of the mechanical and hydraulic behaviour of prestressed concrete-lined pressure tunnels situated above the groundwater level and embedded deep in a rock mass. The rock mass supporting the tunnel was distinguished based on either an elastic isotropic, elasto-plastic isotropic, and elastic transversely isotropic material.

On the one hand, analytical approaches to determine stresses and deformations around the tunnel do exist; however, their applicability is limited to a certain extent for the conditions in the tunnel. On the other hand, numerical models have been a powerful alternative to solve engineering problems, but they are rarely calibrated due to lack of in-situ measurements. This research illustrates that there is a global coherence between the analytical and numerical results. Some of the numerical results even have shown to converge toward the closed-form solutions with great accuracy. However, remaining challenges in modelling the behaviour of prestressed concrete-lined pressure tunnels include:

1. In view of continuous contact throughout the system, the simulation of creep, shrinkage and temperature changes were not considered in the model. The effects of strain losses due creep, shrinkage and temperature changes was considered indirectly during gap grouting by adopting the effective grouting pressure obtained by using the modified load-line diagram, which considers the strain losses due to creep, shrinkage and temperature changes.

2. For cases of pressure tunnels in anisotropic rock formations, the rock mass was assumed as a transversely isotropic material. Thereby, the rock mass as a whole is continuous and the influence of stratification planes characteristics is taken into account by incorporating different deformability properties at directions parallel and perpendicular to the stratification planes. This approach is acceptable as long as there is no separation along the planes of transverse isotropy. If the behaviour of the rock mass supporting the tunnel is controlled by the persistent discontinuities in the rock mass, a new approach considering the whole discontinuities network in the rock mass including dominant and secondary discontinuities needs to be developed so as to obtain a more realistic result.

3. An important process that is still missing in the modelling of cracking in pressure tunnel linings, is updating the new permeability of concrete lining as soon as longitudinal cracks occur in the final lining. Like in many of finite element codes, DIANA requires a permeability coefficient as one of the input parameters so as to generate seepage. In the future, this process needs to be considered in view of improving the current state of the art numerical models.

4. Up to the present, it has been rarely possible to determine exactly the pressure acting behind the final lining when the pressure tunnel is put into operation. Further research in this direction is encouraged to avoid uncontrollable estimates of stresses in most cases.

References

Amberg, F. (1997). *Vereina Tunnel, Project Concept and Realisation*, Tunnels for People: World Tunnel Congress' 97, Vienna, Austria: Proceedings, 23rd General Assembly of the International Tunnelling Association, 12-17 April 1997. AA Balkema, pp. 267-271.

Amorim, D.L.d.F., Proença, S.P., Flórez-López, J. (2014). *Simplified Modeling of Cracking in Concrete: Application in Tunnel Linings*. Engineering Structures, 70:23-35.

Anagnostou, G., Kovari, K. (1993). Significant Parameters in Elastoplastic Analysis of Underground Openings. Journal of Geotechnical Engineering, 119(3):401-419.

Barla, G., Bonini, M., Semeraro, M. (2011). *Analysis of the Behaviour of a Yield-Control Support System in Squeezing Rock*. Tunnelling and Underground Space Technology, 26(1):146-154.

Barton, N. (2004). *The Why's and How's of High Pressure Grouting*. Tunnels and Tunnelling International, Part I, September 2004, pp. 28-30.

Barton, N., Buen, B., Roald, S. (2001). *Strengthening the Case for Grouting*. Tunnels and Tunnelling International, Part I: Vol. 33 (12): 34-36 and Part II: Vol. 34 (1): 37-39.

Bathe, K.J. (1982). *Finite Element Procedures in Engineering Analysis*. Prentice Hall, Englewood Cliffs, 1982.

Beeby, A.W., Narayanan, R. (1995). *Designers' Handbook to Eurocode 2: 1. Design of Concrete Structures*. Telford.

Bian, K., Xiao, M., Chen, J. (2009). *Study on Coupled Seepage and Stress Fields in the Concrete Lining of the Underground Pipe with High Water Pressure*. Tunnelling and Underground Space Technology, 24(3):287-295.

Bobet, A. (2011). *Lined Circular Tunnels in Elastic Transversely Anisotropic Rock at Depth*. Rock Mechanics and Rock Engineering, 44(2):149-167.

Bonini, M., Lancellotta, G., Barla, G. (2013). *State of Stress in Tunnel Lining in Squeezing Rock Conditions*. Rock Mechanics and Rock Engineering:1-7.

Bouvard, M. (1975). *Les Fuites Des Galeries en Charge en Terrain Sec. Rôle Du Revêtement, Des Injections, Du Terrain*. La Houille Blanche, (4):255-265

Bouvard, M., Niquet, J. (1980). *Ecoulement Transitoires Dans Les Massifs Autour D'une Galerie en Charge* La Houille Blanche, (3):161-168.

Bouvard, M., Pinto, N. (1969). *Amenagement Capivari - Cachoeira. Etude du Puits en Charge*. La Houille Blanche, (7):747-760.

Broch, E. (1982). *The Development of Unlined Pressure Shafts and Tunnels in Norway*, Proc. ISRM International Symposium on Rock Mechanics: Caverns and Pressure Shafts. A.A. Balkema, Aachen, Germany, pp. 545-554.

Cantieni, L., Anagnostou, G. (2011). *On a Paradox of Elasto-Plastic Tunnel Analysis*. Rock Mechanics and Rock Engineering, 44(2):129-147.

Carranza-Torres, C. (2004). *Elasto-Plastic Solution of Tunnel Problems using the Generalized form of the Hoek-Brown Failure Criterion*. International Journal of Rock Mechanics and Mining Sciences, 41, Supplement 1(0):629-639.

Carranza-Torres, C., Fairhurst, C. (1999). *The Elasto-Plastic Response of Underground Excavations in Rock Masses that Satisfy the Hoek–Brown Failure Criterion*. International Journal of Rock Mechanics and Mining Sciences, 36(6):777-809.

Carranza-Torres, C., Fairhurst, C. (2000a). *Application of the Convergence-Confinement Method of Tunnel Design to Rock Masses that Satisfy the Hoek-Brown Failure Criterion.* Tunnelling and Underground Space Technology, 15(2):187-213.

Carranza-Torres, C., Fairhurst, C. (2000b). *Some Consequences of Inelastic Rock-Mass Deformation on the Tunnel Support Loads Predicted by the Einstein and Schwartz Design Approach*, Trends in Rock Mechanics, pp. 16-49.

Clausen, J., Damkilde, L. (2008). *An Exact Implementation of the Hoek–Brown Criterion for Elasto-Plastic Finite Element Calculations.* International Journal of Rock Mechanics and Mining Sciences, 45(6):831-847.

Deere, D.U. (1983). *Unique Geotechnical Problems at Some Hydroelectric Projects*, Proceedings of the Seventh Panamerican Conference in Soil Mechanics and Foundation Engineering. Associacao Brasileira de Mecahnica dos Solos, Sao Paulo, Brazil, pp. 865-888.

Deere, D.U., Lombardi, G. (1989). *Lining of Pressure Tunnels and Hydrofracturing Potential*, De Mello Volume, Brasil, pp. 121-128.

Detournay, E., Fairhurst, C. (1987). *Two-Dimensional Elasto-Plastic Analysis of a Long, Cylindrical Cavity Under Non-Hydrostatic Loading.* International Journal of Rock Mechanics and Mining Sciences, 24(4):197-211.

Dezi, L., Leoni, G., Tarantino, A.M. (1998). *Creep and Shrinkage Analysis of Composite Beams.* Progress in Structural Engineering and Materials, 1(2):170-177.

DIANA. (2012). *User's Manual - Release 9.4.4.*

Eberhardt, E. (2001). *Numerical Modelling of Three-Dimension Stress Rotation Ahead of an Advancing Tunnel Face.* International Journal of Rock Mechanics and Mining Sciences, 38(4):499-518.

Einstein, H.H., Schwartz, C.W. (1979). *Simplified Analysis for Tunnel Supports.* Journal of the Geotechnical Engineering Division, 105(4):499-518.

Feenstra, P.H. (1993). *Computational Aspects of Biaxial Stress in Plain and Reinforced Concrete.*

Fernandez, G. (1994). *Behavior of Pressure Tunnels and Guidelines for Liner Design.* Journal of Geotechnical Engineering, ASCE, 120(10):1768-1789.

Fortsakis, P., Nikas, K., Marinos, V., Marinos, P. (2012). *Anisotropic Behaviour of Stratified Rock Masses in Tunnelling.* Engineering Geology, 141:74-83.

Gao, Z., Zhao, J., Yao, Y. (2010). *A Generalized Anisotropic Failure Criterion for Geomaterials.* International Journal of Solids and Structures, 47(22):3166-3185.

González-Nicieza, C., Álvarez-Vigil, A.E., Menéndez-Díaz, A., González-Palacio, C. (2008). *Influence of the Depth and Shape of a Tunnel in the Application of the Convergence Confinement Method.* Tunnelling and Underground Space Technology, 23(1):25-37.

Grunicke, U.H., Ristić, M. (2012). *Prestressed Tunnel Lining – Pushing Traditional Concepts to New Frontiers / Neue Grenzen für Passiv Vorgespannte Druckstollenauskleidungen.* Geomechanics and Tunnelling, 5(5):503-516.

Hefny, A.M., Lo, K.Y. (1999). *Analytical Solutions for Stresses and Displacements around Tunnels Driven in Cross-Anisotropic Rocks.* International Journal for Numerical and Analytical Methods in Geomechanics, 23(2):161-177.

Hendron, A.J., Fernandez, G., Lenzini, P.A., Hendron, M.A. (1989). *Design of Pressure Tunnels*, The Art and Science of Geotechnical Engineering and the Dawn of the Tweenty First Century. In: Cording, E. J., Hall, W. J., Haltiwanger, J. D., Hendron, A. J., Mesri, G. Prentice Hall, pp. 161-192.

Hoek, E. (2000). *Practical Rock Engineering.* Internet Reference: www.rocscience.com.

Hoek, E., Brown, E.T. (1980a). *Empirical Strength Criterion for Rock Masses.* Journal of Geotechnical and Geoenvironmental Engineering, 106(ASCE 15715).

Hoek, E., Brown, E.T. (1980b). *Underground Excavation in Rock.* The Institution of Mining and Metallurgy, London.

Hoek, E., Brown, E.T. (1997). *Practical Estimates of Rock Mass Strength.* International Journal of Rock Mechanics and Mining Sciences, 34(8):1165-1186.

Hoek, E., Carranza-Torres, C., Corkum, B. (2002). *Hoek-Brown Failure Criterion - 2002 Edition*, In: Proceedings of the Fifth North American Rock Mechanics Symposium, Toronto, Canada, pp. 267-273.

Hoek, E., Marinos, P. (2000). *Predicting Tunnel Squeezing Problems in Weak Heterogeneous Rock Masses.* Tunnels and Tunnelling International, 32(11):45-51.

Hudson, J.A., Harrison, J.P. (2001). *Engineering Rock Mechanics.* Part 2: Illustrative Worked Examples, 2. Elsevier Science.

Jaeger, C. (1955). *Present Trends in the Design of Pressure Tunnels and Shafts for Underground Hydroelectric Power Stations*, ICE Proceedings. Thomas Telford, pp. 116-174.

Jing, L. (2003). *A Review of Techniques, Advances and Outstanding Issues in Numerical Modelling for Rock Mechanics and Rock Engineering.* International Journal of Rock Mechanics and Mining Sciences, 40(3):283-353.

Jing, L., Hudson, J.A. (2002). *Numerical Methods in Rock Mechanics.* International Journal of Rock Mechanics and Mining Sciences, 39(4):409-427.

Kastner, H. (1962). *Statik des Tunnel- und Stollenbaues.* Springer-Verlag, Berlin, Gottingen, Heidelberg.

Kieser, A. (1960). *Druckstollenbau.* Springer-Verlag, Vienna, Austria.

Kolymbas, D., Wagner, P., Blioumi, A. (2012). *Cavity Expansion in Cross-Anisotropic Rock.* International Journal for Numerical and Analytical Methods in Geomechanics, 36(2):128-139.

Lauffer, H. (1968). Vorspanninjektionen für Druckstollen. In: Müller, L. (Ed.), *Aktuelle Probleme der Geomechanik und Deren Theoretische Anwendung / Acute Problems of Geomechanics and Their Theoretical Applications.* Felsmechanik und Ingenieurgeologie / Rock Mechanics and Engineering Geology. Springer Vienna, pp. 207-208.

Lauffer, H., Seeber, G. (1961). *Design and Control of Linings of Pressure Tunnels and Shafts, Based on Measurements of the Deformability of the Rock*, 7th ICOLD Congress, Rome, Italy, pp. 679-709.

Lu, A.-Z., Xu, G.-S., Zhang, L.-Q. (2011). *Optimum Design Method for Double-Layer Thick-Walled Concrete Cylinder with Different Modulus.* Materials and Structures, 44(5):923-928.

Manh, H.T., Sulem, J., Subrin, D. (2014). *A Closed-Form Solution for Tunnels with Arbitrary Cross Section Excavated in Elastic Anisotropic Ground.* Rock Mechanics and Rock Engineering:1-12.

Marence, M. (1996). *Finite Element Modelling of Pressure Tunnel*, Proceedings of the 2nd ECCOMAS Conference. In: Computational Methods in Applied Sciences '96. J. A. Desideri, C. Hirsch, P. Le Tallec, E. Onate, M. Pandolfi, J. Periaux, E. Stein. John Wiley & Sons Ltd, pp. 211-217.

Marence, M. (2008). *Numerical Modelling and Design of Pressure Tunnels*, Proceedings of The International Conference HYDRO 2008, Ljubljana, Slovenia.

Marence, M., Oberladstätter, A. (2005). *Design of Pressure Tunnels of Ermenek Hydropower Plant*, Proceedings of The 31st ITA-AITES World Tunnel Congress, Istanbul, Turkey, pp. 53-58.

Marinos, V., Marinos, P., Hoek, E. (2005). *The Geological Strength Index: Applications and Limitations*. Bulletin of Engineering Geology and the Environment, 64(1):55-65.

Matt, P., Thurnherr, F., Uherkovich, I. (1978). *Prestressed Concrete Pressure Tunnels*. Water Power & Dam Construction, (May 1978):23-31.

ÖNORM, B. (2001). *4700: Stahlbetontragwerke*. EUROCODE-nahe Berechnung, Bemessung und Konstruktive Durchbildung, 1.

Oreste, P.P. (2003). *Analysis of Structural Interaction in Tunnels Using the Covergence Confinement Approach*. Tunnelling and Underground Space Technology, 18(4):347-363.

Panet, M., Givet, P.D.C.O., Guilloux, A., Duc, J.L.D.G.T.M.N.M., Piraud, J., Wong, H.T.S.D.H. (2001). *The Convergence Confinement Method*. AFTES–Recommendations des Groupes de Travait.

Panet, M., Guenot, A. (1982). *Analysis of Convergence Behind The Face of a Tunnel*, Proceedings, International Symposium Tunnelling'82, The Institution of Mining and Metallurgy. London, pp. 197-204.

Panthi, K.K., Nilsen, B. (2010). *Uncertainty Analysis for Assessing Leakage through Water Tunnels: A Case from Nepal Himalaya*. Rock Mechanics and Rock Engineering, 43(5):629-639.

PCA. (1979). *Design and Control of Concrete Mixtures*. Portland Cement Association.

RocLab. (2002). *Rocscience Inc*. Toronto, Ontario.

Schaedlich, B., Schweiger, H.F. (2014). *A New Constitutive Model for Shotcrete*. Numerical Methods in Geotechnical Engineering, 1:103.

Schleiss, A. (1988). *Design Criteria Applied for the Lower Pressure Tunnel of the North Fork Stanislaus River Hydroelectric Project in California*. Rock Mechanics and Rock Engineering, 21(3):161-181.

Schleiss, A. (2013). *Competitive Pumped-Storage Projects with Vertical Pressure Shafts without Steel Linings*. Geomechanics and Tunnelling, 6(5):456-463.

Schleiss, A.J. (1986a). *Bemessung von Druckstollen*. Teil II: Einfluss der Sickerströmung in Betonauskleidung und Fels, Mechanisch-Hydraulische Wechselwirkungen, Bemessungskriterien, Mitteilung der Versuchsanstalt für Wasserbau, Hydrologie und Glaziologie, No. 86, ETH Zürich, Switzerland.

Schleiss, A.J. (1986b). *Design of Pervious Pressure Tunnels*. Water Power & Dam Construction, 38(5):21-26, 29.

Schleiss, A.J. (1987). *Design Criteria for Pervious and Unlined Pressure Tunnels*, International Conference on Hydropower. Underground Hydropower Plants, Oslo, Norway, pp. 667-677.

Schleiss, A.J. (1997a). *Design of Concrete Linings of Pressure Tunnels and Shafts for External Water Pressure*, Tunnelling ASIA, New Delhi, India, pp. 291-300.

Schleiss, A.J. (1997b). *Design of Reinforced Concrete Linings of Pressure Tunnels and Shafts*. Hydropower & Dams, 4(3):88-94.

Schleiss, A.J., Manso, P.A. (2012). *Design of Pressure Relief Valves for Protection of Steel-Lined Pressure Shafts and Tunnels Against Buckling During Emptying*. Rock Mechanics and Rock Engineering, 45(1):11-20.

Schürch, R., Anagnostou, G. (2012). *The Applicability of the Ground Response Curve to Tunnelling Problems that Violate Rotational Symmetry*. Rock Mechanics and Rock Engineering, 45(1):1-10.

Schütz, R., Potts, D., Zdravkovic, L. (2011). *Advanced Constitutive Modelling of Shotcrete: Model Formulation and Calibration*. Computers and Geotechnics, 38(6):834-845.

Schwarz, J. (1985). *Druckstollen und Druckschächte*: Bemessung und Konstruktion. Techn. Univ. München, Lehrstuhl für Wasserbau und Wassermengenwirtschaft im Inst. für Bauingenieurwesen IV.

Seeber, G. (1982). *Neue Entwicklungen im Druckstollenbau*, ISRM International Symposium.

Seeber, G. (1984). *Recent Developments in the Design and Construction of Power Conduits for Storage Power Stations*, Idraulica del Territorio Montano, Bressanone, Italy, pp. 177-204.

Seeber, G. (1985a). *Power Conduits for High-Head Plants, Part One*. International Water Power & Dam Construction, 37(6):50-54.

Seeber, G. (1985b). *Power Conduits for High-Head Plants, Part Two*. International Water Power and Dam Construction, 37(7):95-98.

Seeber, G. (1999). *Druckstollen und Druckschächte*: Bemessung-Konstruktion-Ausführung. Enke im Georg Thieme Verlag.

Serrano, A., Olalla, C., Reig, I. (2011). *Convergence of Circular Tunnels in Elastoplastic Rock Masses with Non-Linear Failure Criteria and Non-Associated Flow Laws*. International Journal of Rock Mechanics and Mining Sciences, 48(6):878-887.

Sharan, S.K. (2005). *Exact and Approximate Solutions for Displacements around Circular Openings in Elastic-Brittle-Plastic Hoek-Brown Rock*. International Journal of Rock Mechanics and Mining Sciences, 42(4):542-549.

Sharp, J.C., Gonano, L.P. (1982). *Rock Engineering Aspects of the Concrete Lined Pressure Tunnels of the Drakensberg Pumped Storage Scheme*, ISRM International Symposium. International Society for Rock Mechanics.

Simanjuntak, T.D.Y.F., Marence, M., Mynett, A.E. (2012a). *Towards Improved Safety and Economical Design of Pressure Tunnels*, ITA-AITES World Tunnel Congress & 38th General Assembly (WTC 2012), Bangkok, Thailand.

Simanjuntak, T.D.Y.F., Marence, M., Mynett, A.E., Schleiss, A.J. (2013). *Mechanical-Hydraulic Interaction in the Cracking Process of Pressure Tunnel Linings*. Hydropower & Dams, 20(5):112-119.

Simanjuntak, T.D.Y.F., Marence, M., Mynett, A.E., Schleiss, A.J. (2014a). *Effects of Rock Mass Anisotropy on Deformations and Stresses around Tunnels During Excavation*, The 82nd Annual Meeting of ICOLD, International Symposium on Dams in a Global Environmental Challenges, Bali Nusa Dua, Indonesia, pp. II-129 - II-136.

Simanjuntak, T.D.Y.F., Marence, M., Mynett, A.E., Schleiss, A.J. (2014b). *Longitudinal Cracks in Pressure Tuunels: Numerical Modelling and Structural Behaviour of Passive Prestressed Concrete Linings*, Numerical Methods in Geotechnical Engineering. CRC Press, pp. 871-875.

Simanjuntak, T.D.Y.F., Marence, M., Mynett, A.E., Schleiss, A.J. (2014c). *Pressure Tunnels in Non-Uniform In Situ Stress Conditions*. Tunnelling and Underground Space Technology, 42(0):227-236.

Simanjuntak, T.D.Y.F., Marence, M., Schleiss, A.J., Mynett, A.E. (2012b). *Design of Pressure Tunnels Using a Finite Element Model*. Hydropower & Dams, 19(5):98-105.

Stematiu, D., Lacatus, F., Popescu, D. (1982). *A Finite Element Model for Excavation, Lining and Lining Prestressing of Water Power Plant Tunnels*, Proc. ISRM

International Symposium on Rock Mechanics: Caverns and Pressure Shafts. A.A. Balkema, Aachen, Germany, pp. 735-748.

Stini, J. (1950). Tunnelbaugeologie. Wien. Julius Springer-Verlag, 366.

Swoboda, G., Marence, M., Mader, I. (1993). *Finite Element Modelling of Tunnel Excavation.* International Journal for Engineering Modelling, 6(1-4):51-63.

Timoshenko, S.P., Goodier, J.N., Abramson, H.N. (1970). *Theory of Elasticity.* Journal of Applied Mechanics, 37:888.

Tonon, F. (2004). *Does Elastic Anisotropy Significantly Affect a Tunnel's Plane Strain Behavior?* Transportation Research Record: Journal of the Transportation Research Board, 1868(1):156-168.

Tonon, F., Amadei, B. (2003). *Stresses in Anisotropic Rock Masses: An Engineering Perspective Building on Geological Knowledge.* International Journal of Rock Mechanics and Mining Sciences, 40(7):1099-1120.

Vigl, A., Gerstner, R. (2009). *Grouting in Pressure Tunnel Construction.* Geomechanics and Tunnelling, 2(5):439-446.

Vlachopoulos, N., Diederichs, M.S. (2009). *Improved Longitudinal Displacement Profiles for Convergence Confinement Analysis of Deep Tunnels.* Rock Mechanics and Rock Engineering, 42(2):131-146.

Vu, T., Sulem, J., Subrin, D., Monin, N. (2013). *Semi-Analytical Solution for Stresses and Displacements in a Tunnel Excavated in Transversely Isotropic Formation with Non-Linear Behavior.* Rock Mechanics and Rock Engineering, 46(2):213-229.

Wan, R.G. (1992). *Implicit Integration Algorithm for Hoek-Brown Elastic-Plastic Model.* Computers and Geotechnics, 14(3):149-177.

Wang, Y. (1996). *Ground Response of Circular Tunnel in Poorly Consolidated Rock.* Journal of Geotechnical Engineering, 122(9):703-708.

Wannenmacher, H., Krenn, H., Komma, N., Bauert, M., Grunicke, U. (2012). A Case Study of the Niagara Tunnel Facility Project - Technical and Economical Aspects of Passive Prestressed Pressure Tunnels, Swiss Tunnel Congress 2012, Lucerne, Switzerland.

Wittke, W. (1990). *Rock Mechanics: Theory and Applications, with Case Histories.* Springer-Verlag.

Zienkiewicz, O.C., Morice, P. (1971). *The Finite Element Method in Engineering Science,* 1977. McGraw-hill London.

Zienkiewicz, O.L. (1958). *Effect of Pore Pressure on Stress Distribution in Some Porous Elastic Solids.* Water Power & Dam Construction, (January).

List of Figures

List of Tables

Acknowledgements

And finally here it is. The final part that reminds me of the research carried out in four and a half years at UNESCO-IHE in Delft. No doubt, this research would not have been possible without the help and support from the kind people around me, to only some of whom it is possible to give particular mention here.

First and foremost, I thank the sponsor: Verbund Hydropower AG in Austria for the financial support throughout the years. I am thankful to Mr. Franz Gappmaier for the comments and discussions during my visit to Austria.

My promotor, Prof. dr. ir. A.E. Mynett. Dear Prof. Arthur, you always believed in my work. I am very grateful for your positive attitude, advice and encouragement, which gave me confident especially during difficult times when things did not go as smoothly as hoped.

My co-promotor, Dr. M. Marence. Dear Miro, this research started in a corner of IHE canteen. From you, I have learned the skill of critical thinking. Thank you for your faith, guidance and continuous support. I really enjoyed the opportunity to watch and learn from your knowledge and experience.

This research would not have been completed without the wise counsel of Prof. dr. A.J. Schleiss. Dear Prof. Schleiss, you cared so much about my research. You always responded to my queries promptly. Thank you for your time and advice. It was my great pleasure to have discussions with you at EPFL in Lausanne. You are an inspiring professor and I owe a lot of my success to your input, particularly in terms of publishing papers.

For this dissertation, I would like to express my deep and sincere gratitude to the committee members: Prof. dr. S. Uhlenbrook, Prof. dr. ir. J.A. Roelvink, Dr. ir. D.J.M. Ngan-Tillard, Dr. R. Kohler and Prof. dr. ir. G.S. Stelling, for spending the time on reading and offering constructive comments and recommendations on the draft version.

Many thanks to the people at UNESCO-IHE including Gordon de Wit, Willem van Nievelt, and Guy Beaujot for helping me with the software I need. Tonneke Morgenstond, for your kind help and providing me with a work place overlooking the Delft canal. Suryadi, Sylvia van Opdorp-Stijlen, Anique Karsten, Peter Stroo, Ed Gerritsen van der Hoop, Gerda de Gijsel, Jos Bult, Martine Roebroeks-Nahon and Jolanda Boots for all you have done to support me.

I would like to thank Gladys for her time and efforts to look up and send me books and journal papers I need from Berlin. Dejen, Girma, Yenesewu, Linh, Mona, Yuli, Selvi, Off, Anwar, Isnaeni, Sony, Gabriela, Clara and Tan, thank you for all the good time we spent together. I will miss the fun and laugh each day we had during lunch time. My special thanks to Marti, Nadine, Uli, Nova, Waya, and Hendra for being compassionate. May our friendship last forever.

Dear paranymphs, what an honour to have both of you standing next to me during the defence. Laurens, I like to have the opportunity of calling and talking to you whenever I want. I will always remember that we once organized the annual PhD Symposium together. Leo, I can always turn to you for advice even our research topics are different. All the moments we shared, the happy and sad ones made our bound this strong. You did surprise me with your quick decision when I asked you for this job. Thank you for all the tea and coffee drinking times.

To my current employer PLAXIS, I am very grateful to Jan-Willem Koutstaal and Erwin Beernink for trusting me, hiring me and giving me the opportunity to grow up at the Sales and Marketing Department. I would like to thank Annelies, Janine, Karin, Szilvia, Arijana, Bianca, Michael, Vincent and Yannick for creating such a pleasant working atmosphere. My special thanks to Judi, for his support and help in designing the layout of the thesis cover. I would also like to thank my colleagues from the Research Department including Ed, Luc, Micha, and Thomas for the nice conversation and fun during lunch time.

Then, I would like to express my deepest gratitude to the most important people in my life.

My best friends and jazz lovers: Nadia, Maia, Caesar, Heidi and Welling. As four of you already left The Netherlands, listening to jazz music will never be the same without you. Thanks for making so many ordinary moments, extraordinary. Our memories will last me a lifetime.

Shah, what can I say? I cannot thank you enough for everything you have done for me, especially in the last four years. You are like family to me and I look forward to witnessing all the things you will accomplish.

Mijn lieve schoonfamilie: Omi en Opa Tooi, Tineke, Peter, Jasper, Tibby, Esther en Aurélien. Wat een warme en aardige mensen. Ik voel me altijd welkom bij jullie. Hartelijk dank voor jullie steun, aandacht en interesse in wat ik doe en wie ik ben. Ik dank Harm, Elly en Sanne voor het meeleven en meedenken. Het is geweldig om dit moment met jullie te delen.

Tersayang Mama dan Papa, untuk kasih sayang yang tak terhingga. Terima kasih atas segala doa, dukungan dan pengorbanan yang telah diberikan demi tercapainya cita-cita yang Yos impikan. Tiada kata yang mampu mengungkapkan betapa Yos bangga memiliki orang tua seperti kalian.

Liefste Rutger. Wat ben ik erg blij dat we elkaar hebben gevonden. Jouw vertrouwen in mij, hulp en steun bij zoveel dingen rondom mijn promotietraject en proefschrift zijn van onschatbare waarde geweest. Ik weet dat je net zo blij en opgelucht bent als ik over het feit dat dit onderzoek succesvol is afgerond. Alle leuke en bijzondere dingen die we afgelopen jaren gedaan en meegemaakt hebben zijn belangrijk voor mij geweest. Ik kijk ernaar uit om onze dromen na te jagen!

Utrecht, March 2015.

About the Author

Yos Simanjuntak was born on 14 July 1978 in Pangkalan Susu, Indonesia. In 2002, he obtained his BSc (Cum Laude) in Civil Engineering from the Institut Teknologi Medan in Indonesia, where he was later on appointed as an Assistant Lecturer. Aside from organizing seminar and workshops at the institute, his activities involved teaching on civil engineering subjects including Hydraulics, Dam Engineering and Concrete Structures. He continued to do so until 2005.

From May to July 2005, he worked as a Program Assistant for the United Nations WHO on the small island of Nias in Sumatra, which, as so many other communities surrounding the Indian Ocean, had suffered dearly from the devastating Tsunami during Christmas 2004. It was his responsibility to coordinate health efforts of local and international humanitarian organizations as well as surveys on healthcare facilities in the disaster affected areas. It was also during his stay on this island when he informed that he was granted a Nuffic scholarship, which enabled him to deepen and broaden his interests in hydraulics. In 2007, he received an MSc (with Distinction) in Hydraulic Engineering at UNESCO-IHE in Delft, The Netherlands.

In November 2010, he started his PhD research at UNESCO-IHE/TU Delft focusing on the design of prestressed concrete-lined pressure tunnels. Besides doing research, he organized PhD symposium in a joint effort with fellow PhD candidates in 2013. He has guided several MSc students and given lectures on the topic of the Design of Settling Basins for the short course Small Hydropower Developments at UNESCO-IHE until 2014.

Currently, he works at Plaxis B.V in Delft as a Sales Support Engineer, whose responsibilities are, among others, providing advice to new and existing customers worldwide on the various PLAXIS products. As the name suggests, PLAXIS provides solutions based on the finite element method dedicated to civil and geotechnical engineering.

Publications

Journal Papers

1. Simanjuntak, T.D.Y.F., Marence, M., Schleiss, A.J., Mynett, A.E. *The Interplay of In-Situ Stress Ratio and Transverse Isotropy in the Rock Mass on Prestressed Concrete-Lined Pressure Tunnels.* Tunnelling and Underground Space Technology (Under Review).
2. Simanjuntak, T.D.Y.F., Marence, M., Mynett, A.E. Schleiss, A.J. (2014). *Pressure Tunnels in Non-Uniform In Situ Stress Conditions.* Tunnelling and Underground Space Technology, 42, 227-236.

3. Simanjuntak, T.D.Y.F., Marence, M., Mynett, A.E. Schleiss, A.J. (2013). *Mechanical-Hydraulic Interaction in the Cracking Process of Pressure Tunnel Linings*. Hydropower & Dams, 20(5): 112-119.
4. Simanjuntak, T.D.Y.F., Marence, M., Schleiss, A.J., Mynett, A.E. (2012). *Design of Pressure Tunnels Using a Finite Element Model*. Hydropower & Dams, 19(5): 98-105.

Conference Papers

1. Simanjuntak, T.D.Y.F., Marence, M., Mynett, A.E. (2012). *Towards Improved Safety and Economical Design of Pressure Tunnels*. ITA-AITES World Tunnel Congress & 38th General Assembly (WTC 2012), Bangkok, Thailand. ISBN 978-974-7197-78-5.
2. Simanjuntak, T.D.Y.F., Marence, M., Mynett, A.E., Schleiss, A.J. (2014). *Effects of Rock Mass Anisotropy on Deformations and Stresses around Tunnels during Excavation*. The 82nd Annual Meeting of ICOLD, International Symposium on Dams in a Global Environmental Challenges. Bali, Indonesia. pp. II-129 – II-136.
3. Simanjuntak, T.D.Y.F., Marence, M., Mynett, A.E., Schleiss, A.J. (2014). *Longitudinal Cracks in Pressure Tunnels: Numerical Modelling and Structural Behaviour of Passive Prestressed Concrete Linings*. Numerical Methods in Geotechnical Engineering. CRC Press, pp. 871-875. ISBN 978-1-138-00146-6.

∞